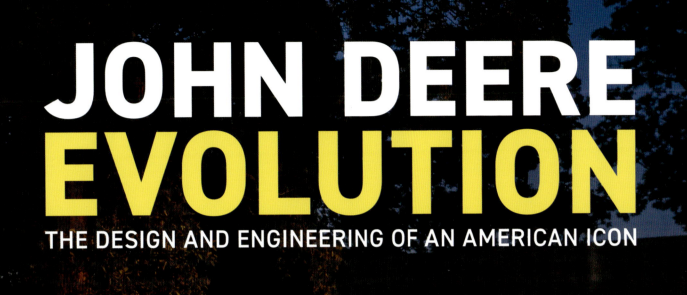

JOHN DEERE
EVOLUTION
THE DESIGN AND ENGINEERING OF AN AMERICAN ICON

LEE KLANCHER

OCTANE
PRESS

CONTENTS

TEN YEARS IN THE MAKING

"The best way out is always through."
—Robert Frost

HANGAR PHOTO STUDIO
This temporary studio was built in a B-52 hangar at the long-abandoned Chanute Air Force Base in Rantoul, Illinois.

This book is the result of a decade-long odyssey into the world of John Deere machines. The imagery is the central appeal of this work, and you'll find a blend of studio and outdoor images of important John Deere machines and archival images of production as well as prototype machines. Making the images, particularly in studio, was one of the most complex photographic endeavors I've undertaken, requiring large amounts of equipment, crew, and resources. And finding the images as well as interviewing the engineers and designers who built John Deere tractors required countless hours of research and travel.

The first major image creation project of the book was working with Bruce and Walter Keller to build a photo studio to capture some of their rare John Deere tractors. They have more than six hundred machines and, more significantly, a large number of serial number one and rare examples.

I returned to the studio in 2019 at the Half Century of Farm Progress Show, where I built another studio in an old hangar on Chanute Air Force Base to capture images of the fantastic machines in the show. We changed techniques a bit, opting for black backgrounds and a natural rather than white floor. The studio creates very pure images. All you have to work with is the machine itself and lush, velvety light. It is a harsh but highly rewarding environment.

The challenges in this book were significant. On the way home from the original Keller studio shoot, my Audi's engine gave out, stranding me in Dallas (where my wife would contend, I should have left that car for dead!). Snow continued to play a role in this book when I planned an upper Midwest book trip in October 2019. High winds, sub-zero weather, and several inches of snow made the roads impassable and froze my camper's water systems solid. Once the wind died down and the salt trucks took the edge off the icy roads, I was able to drag the camper out of the snow and point my rig south. The camper and I thawed out in central Arkansas, nursing the pain of lost photo shoots and media appearances.

Finding historic imagery for the book also was a significant challenge. I was able to get a handful of key images from the John Deere archives, and needed to turn elsewhere to find fresh images.

The Cooper Hewitt Smithsonian Design Museum in New York houses the archives of the Henry Dreyfuss Associates, which designed the exterior of Deere machines from the 1930s to recent times. The company would maintain a close relationship with Deere leadership for more than seven decades, and their work was critical to the brand's ascent.

Frequent visits to those archives yielded a treasure trove of imagery lost to history for several decades. I spent many exciting days at the facility viewing previously uncatalogued material that had not been seen in thirty years. I also found images in the Library of Congress, Smithsonian Collections, National Archives, Standard Oil Collection of New Jersey, and more.

We also were gifted with great images that came from legendary industrial designer Chuck Pelly's collection, taking you back to the creation of the 30 series, as well as Deere engineer Jim Mannisto, Mecum Auctions, and Machinery Pete.

While the images are the star of the show in this book, I was also fortunate to make contact with many people who played significant roles in creating John Deere tractors. Such interviews are always enlightening, and capturing their stories has been one of the most satisfying parts of creating this book.

Lee Klancher

January 2021

A CALL TO ARMS

In December 1909, recently anointed Deere CEO William Butterworth received a telephone call that would shape the arc of his family's company for decades.

Butterworth had stepped in when the innovative second son of the company namesake, Charles Deere, passed away in 1907. Charles had taken on leadership of the company when he was only twenty-one years old. At times bucking the ideas and impulses of his father, John Deere, Charles had overseen the transformation of Deere from a regional plow manufacturer to a major American implement manufacturer.

Butterworth was Charles's son-in-law and had served as an informal second-in-command to Charles. He was the heir apparent, but his reviews were not entirely favorable, inside and outside of Deere.

Butterworth stepped in at a critical and difficult time. In 1907, an era of massive mergers in American business was coming to a close. Notable mergers that consolidated businesses to create one dominant company include those that created U.S. Steel, General Electric (GE), and the International Harvester Company (IH). Such companies were able to completely dominate industries, at times putting more than 70 percent of the competition under one roof.

Particularly in the early years, IH used its already unfair advantage in unscrupulous ways. Dealerships owned by IH competed against each other in the same towns, and companies were purchased without knowing the buyer was IH. International Harvester would later be prosecuted and found guilty in court and forced to break off and sell portions of its business. That suit wasn't filed until 1912, however, and would continue to be debated in court until 1926.

Competing against International Harvester in the early 1900s was tough sledding at best, and the addition of a brutal economic downturn in 1907 put

Butterworth's back against the wall. One of the risky undertakings in front of him was developing a new line of Deere harvesters. Deere's implement line didn't include them, and the dominant maker was International Harvester.

Word leaked to the public that Deere was building a new harvester for the Canadian market. It also came to light that Butterworth was trying to recruit a high-placed executive, McAllister, from International Harvester. When Cyrus McCormick, the president of International Harvester, discovered this, he picked up the telephone and called Butterworth.

According to a recount of the story by McCormick, he suggested to Butterworth that Deere supply plows to International Harvester and IH, in turn, supply harvesters to Deere. A few weeks later, the two met again at lunch to discuss this option.

On January 6, 1910, the directors of Deere met and unanimously resolved an eleven-point plan to consolidate the company management and branches under the John Deere and/or Deere brand, and acquire their subsidiaries as well as a few key others in order to offer a complete line of agricultural implements. It also stated that Deere would maintain a strong focus on the domestic market and a headquarters firmly planted in Moline, Illinois.

The final resolution was to "in some way, out of all this, get the company into harvesting equipment."

The discussion between McCormick and Butterworth about swapping plows for harvesters was dropped, and the mold was set for John Deere's path in the twentieth century.

Deere would build in-house, pick its battles with care, and stay independent. In a fierce marketplace, facing a competitor who owned more than 80 percent of the market, Deere would go it alone.

WILLIAM BUTTERWORTH

The infamous John Deere leader and tractor detractor shown on November 19, 1929. *Library of Congress*

WEATHERED IRON

Nine decades of wear show on this 1937 Model A.

Lee Klancher

CHAPTER ONE

BLACKSMITHS AND CRAFTSMEN

THE FIRST JOHN DEERE TRACTORS

The work was more of an art—a very fascinating art—than a science. . . . Every detail stood by itself, and had to be learnt either by trial and error or by tradition."
—*George Sturt*, The Wheelwright's Shop (1930)

THE FIRST JOHN DEERE TRACTOR

The first tractor to be put into production by John Deere is the Waterloo Boy.

1892 FROELICH TRACTOR

This machine is considered the first farm tractor and was created by Iowa inventor John Froelich. He helped found the Waterloo Gasoline Traction Company, which was later purchased by Deere.

In 1892, an innovative farmer named John Froelich was frustrated with the limitations of steam engines to power his threshing machine. Much as many of the founders of agricultural companies before and after him did, he took matters into his own hands and went to work in his blacksmith shop to build a solution.

He mounted a gas engine on a hand-built cart, and the funky little contraption powered his thresher efficiently. Froelich is now credited as the developer of the first farm tractor (and his machine has a loose tie to John Deere).

Tractors were radical technology before the turn of the twentieth century. As is often the case with new technologies, these early tractors were developed by independent entrepreneurs with the vision, courage, and temerity to build machines in blacksmiths, university machine shops, and backyard sheds.

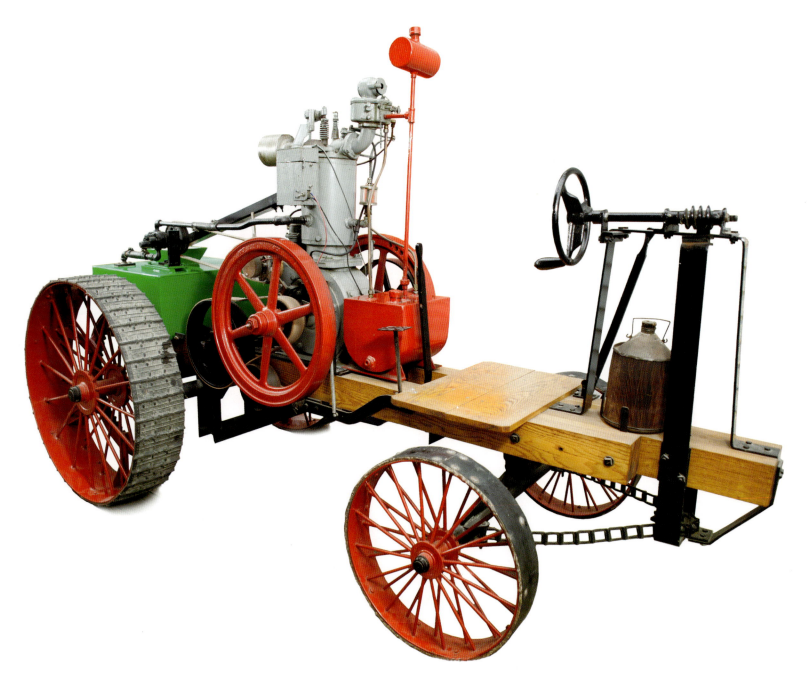

DAIN TRACTOR

In dramatic contrast to the elemental simplicity of the Waterloo Boy is the progressive innovation of the Dain tractor. *John Deere Archives*

John Deere was a power at the time. Since being founded in 1837 by a blacksmith named John Deere, the company had grown to become one of the largest agricultural equipment manufacturers in the world. Deere was faced, however, with a tremendous challenge, as the newly merged International Harvester Company had formed in 1902 with much of the industry under one roof.

The early IH offerings didn't overlap with much of the Deere line of equipment, so Deere's approach was to stick to its lane in the assumption that IH would do the same. This was not the correct approach, however, as IH began rapidly purchasing companies and expanding its markets.

Deere was one of the biggest players in the field at the time, but it only held a single-digit market share

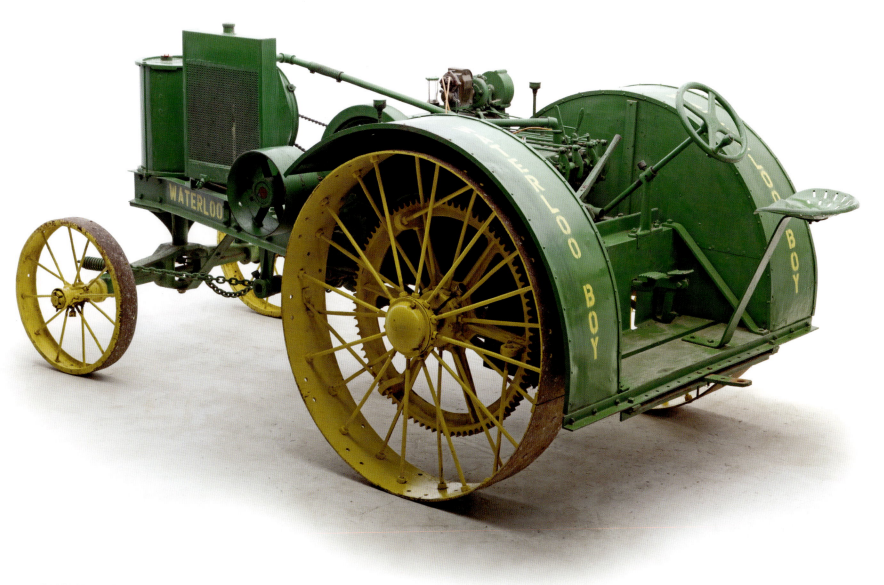

1916 WATERLOO BOY MODEL R

The Waterloo Boy design was refined by Louis Witry in a shed in his backyard. Deere purchased the Waterloo Gasoline Tractor Company in 1918. The Model R engine makes about 12 horsepower at the drawbar, and the machine weighs roughly 6,200 pounds. Top speed is about 2.5 miles per hour. The magneto ignition, gear drive, and fan-cooled radiator cooling system were state of the art in 1916. This Model R uses round spoke wheels and a vertical kerosene tank in the front. The steering system is linked with a chain, which is about as easy to operate as you might expect.

compared to IH's hold, which had been estimated at 70 percent or more of the agricultural equipment industry. Deere was busy consolidating on a number of fronts, combining separate regional entitities into a national organization as well as acquiring companies and expanding its operations. Cash was tight, prompting Charles Deere (the son of the founder, John) to warn up-and-coming company leader William Butterworth to be cautious in a 1905 memo. "Get along with such outfits as you have or such as will," he wrote. "There seems to be no end to outlay of money. Stop it!"

The very earliest tractors were large and expensive, and developed mainly by small builders and independents focused solely on these machines. In this challenging business environment, Deere was cash-strapped and outgunned by the massive IH merger. The company needed to make careful investments, and large farm tractors in the early 1900s required large R&D budgets to reach a narrow market.

While the machines sold for hefty price tags that led to enviable profit margins, the market for this type of tractor was limited to massive farms, farm cooperatives, or thresher operators. The market for these machines peaked in 1912, with thirty-one makers selling 11,500 machines.

Author Barton W. Currie summed up the situation nicely, writing, "It looked like a golden market for the big gas tractor, and the manufacturers went to it. There was an exhilarating boom and then a bang, followed by a sudden collapse. The big gas tractor had been overexploited and oversold."

Deere deftly ducked this trendy development, wisely staying out of the fight for the large tractor market. It distributed the massive Big Four tractor built by the Gas Traction Company, which gave it a large gas tractor in its catalog, and saved the R&D money for more interesting ventures.

The agricultural market's lucrative opportunity at that time was to create a low-price, small tractor useful on most farms, and Deere explored that with several different experimental programs.

In 1912, designer Charles Melvin created an experimental tractor, known as the Melvin Tractor, that was based on a Hackney Motor Plow. Only one prototype was built and it carried "John Deere" script on the side. That tractor performed poorly and was abandoned in 1914.

John Deere engineer Max Sklovsky created several early designs—the A-2, B-2, and D-2. The A-2 and B-2 were built and tested in 1915 and 1916. Both were discontinued, and the D-2 only existed on paper as it was never constructed.

The company's most interesting early tractor development was done by Joseph Dain, who owned a company Deere acquired. His all-wheel-drive Dain Tractor was a progressive design that featured a friction transmission that shifted on the fly and a four-cylinder engine.

On October 31, 1917, inventor Joseph Dain contracted a cold that killed him in two days (and the Spanish Flu broke out a few months later). Deere had authorized building one hundred prototypes shortly before he died, and some of those were built and tested in 1919 before the project was scuttled and buried. Even if Dain had survived, his concept was expensive to build and would have fit into the market as the high-dollar small tractor of choice. That would not be the tractor for the masses sought by Deere and other thought leaders of the time, so it's unlikely that it would have been put into production.

When Henry Ford joined the fray with his Fordson tractor, the entire industry was put on alert. Ford wisely built a low-cost, relatively small machine, sold it in volume to the British government during the end of World War I, and introduced it to the U.S. market in 1917. The loud, cheap, and innovative machine's success transformed the market.

Deere management pushed to find a way to quickly enter the small tractor market, and executive Frank Silloway enthusiastically presented the company the opportunity to purchase the Waterloo Gasoline Traction Engine. The company built the well-regarded Waterloo Boy tractor, and Silloway was enthusiastic

WATERLOO BOY CIRCA 1917

This advertisement shows the paint schemes used on early Waterloo Boy tractors.

Wisconsin Historical Society

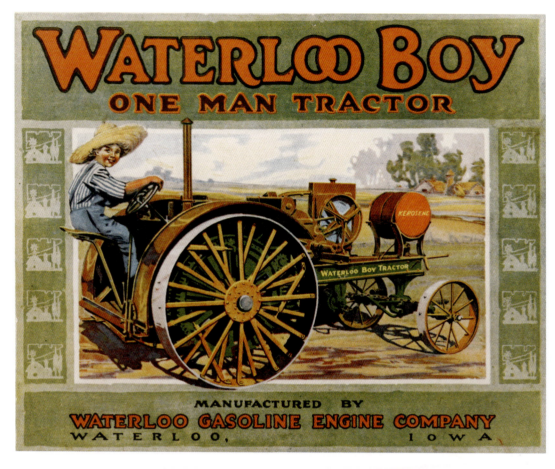

EARLY MODEL DATA

Model	Type	Model Years	Notes	HP	Nebraska Test #
Froelich Tractor	Home Build	1892–1893	Hand-built. Original gone, replicas exist.	16	-
Melvin Tractor	Deere Experimental	1912–1914	One built.	-	-
Waterloo Boy C	Production Model	1913	Kerosene engine.	25	-
Waterloo Boy L (LA)	Production Model	1914	1914 LA chassis became Model R. Kerosene engine with gas start.	16	-
Waterloo Boy R	Production Model	1914–1917	RA-RM. Kerosene engine with gas start.	27	-
Overtime R	Waterloo Boy Export Model	1914–1919	Repainted, rebranded for Great Britain. Kerosene engine with gas start.	27	-
Sklovsky Tractor A-2, B-2, & D-2	Deere Experimental	1915–1916	Note later Sklovsky experimentals were built and tested.	NA	-
Waterloo Boy N	Deere purchased WB March 1918	1917–1924	Kerosene engine with gas start. Last production 10/15/1924.	27	1
Overtime N	Waterloo Boy Export Model	1917–1924	Kerosene engine with gas start. Repainted, rebranded for Great Britain until 03/1918.	27	-
Tractivator	Motor Cultivator	1917	Motor Cultivator, 25 experimentals produced, both one- and two-row versions tested.	-	-
Dain AWD Tractor	Deere Experimental	1918–1919	Deere built roughly 100. Intact original on display at JD Museum in Waterloo, Iowa.	26	-

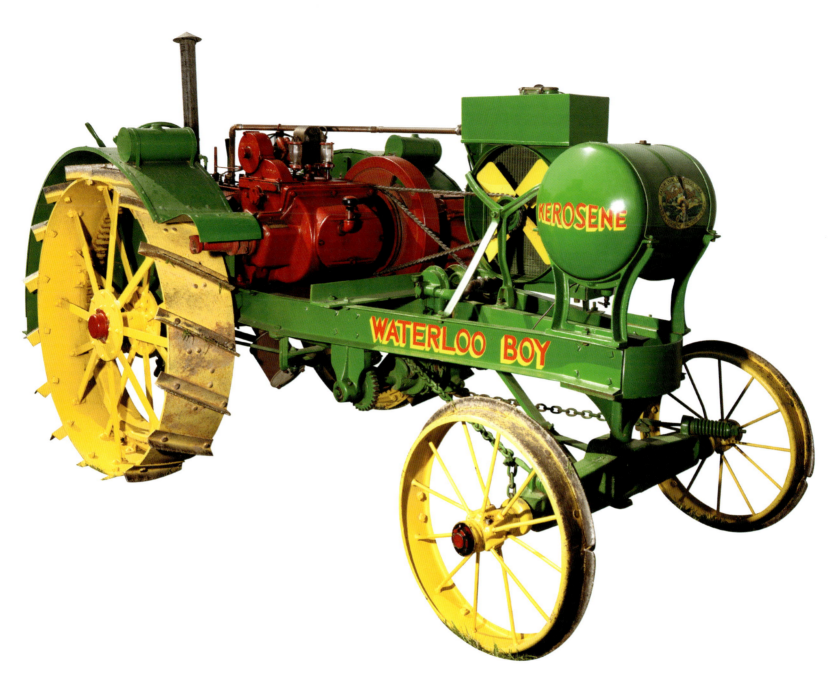

about the machine's quality and the simplicity of the two-cylinder engines.

"The Waterloo tractor is of a type that the average farmer can buy," Silloway wrote. "We should have a satisfactory tractor at a popular price, and not a high-priced tractor for the few. . . . Here we have an opportunity to, overnight, step into first place in the tractor business."

Deere purchased Waterloo Boy in March 1918. While sales, and for that matter the technology on the staid machines, were eclipsed by Ford and other makers, Silloway's enthusiasm was prescient. The Waterloo Boy sold well enough to make the purchase a profitable venture. More significantly, its elemental two-cylinder engine design would define the identity of Deere's tractor line for more than four decades.

WATERLOO BOY

Waterloo Boy tractors are typically painted with a green frame and red engine, as shown here. *Renner Collection / Lee Klancher*

MODEL D (1923-1939)

JOHN DEERE MODEL D

The D was an evolution of the Waterloo Boy design. While the tractor is more compact and uses a smoother unit frame, the basic layout of the components is quite similar.

John Deere Archives

In 1921, the farm economy crashed and demand for expensive high technology—particularly farm tractors—cratered. Ever the firebrand, Ford responded by dropping the price of his Fordson to $395 (and sold thirty-five thousand of them). Despite being ensnared in anti-trust lawsuits, IH had a stranglehold on the ag-equipment market and nearly bottomless cash reserves, so it matched Ford with the rock-bottom price of $900 on its small International 8-16.

The IH and the Ford were both relatively new designs, and the Waterloo Boy had aging technology and dated looks. Sales dropped from more than five thousand Waterloo Boy tractors in 1920 to only seventy-nine tractors the following year.

Facing layoffs and salary cuts, Deere needed to find a way to put a new model on the market. As it happened, it had a new design waiting in the wings. Louis Witry and others at Deere had refined the Waterloo Boy design and constructed a much-improved machine

capable of being built at a competitive price. The new model offered 40 percent greater drawbar pull than the Waterloo Boy, and weight was cut by 33 percent.

Beginning in 1922, Leon Clausen spearheaded the management team's move to transform the Witry design into a new model tractor. Clausen debated if they should develop a more modern four-cylinder engine. However, according to author Wayne G. Broehl Jr., the more modern engine was rejected not by strategy but by market. "The final decision to go ahead with the two-cylinder was made not on technical grounds but on the basis of the short-term embarrassed financial situation," Broehl wrote.

The Model D was introduced as a 1923 model and received by the market with great enthusiasm. The tractor was refined continuously throughout production, and it received a dose of style for the 1939 model year. The Model D continued to be incrementally improved until 1953, making it the longest-lived model in company history.

1923 EARLY SPOKER MODEL D

The first fifty Model D tractors, all produced in 1923, had fabricated front axles and ladder-type radiator sides. *Keller Collection / Lee Klancher*

1926 SPOKER MODEL D

The Model Ds produced in 1923 to about 1925 are known as "spoker" Ds due to their spoked flywheel (models built after December 1925 used a solid flywheel).

Keller Collection / Lee Klancher

1926 SPOKER MODEL D

This 1926 Model D, painted orange at the factory, is one of the first machines the John Deere factory painted anything other than green. Kay Brunner cast wheels are one of this tractor's distinctive features.

1926 SPOKER MODEL D

The reinforced hitch and orange paint suggest this tractor was intended for industrial work. Industrial options were first offered for the D in 1926. The Model DI was first offered as a 1935 model, and the DI was covered in highway yellow paint.

1930 MODEL DX24 EXPERIMENTAL CRAWLER

According to several articles in *Two-Cylinder* magazine, two large production runs of experimental model Ds were built to significantly test and improve the Model D. These "DX" models are well-documented and highly collectible, and known as the "Exhibit A" and "Exhibit B" experimental Model Ds. Author Don MacMillan stated that ninety-six Exhibit A and fifty Exhibit B machines were built. The example shown here is one of a handful of experimental Model DXs fitted by John Deere with tracks, which improved drawbar pull but were reportedly nearly impossible to steer. Less than ten—and possibly as few as three—of these were built, and the machine never saw production. *Keller Collection / Lee Klancher*

MODEL D LINDEMAN CRAWLER

In 1932, dealership owner and inventor Jesse G. Lindeman mounted a set of Best tracks to several Model Ds at his facility in Yakima, Washington. Deere took interest in the machine and had one shipped to the Deere Experimental Farm. Theo Brown tested it on May 2, 1933, and wrote in his diary that drawbar pull was superior to wheeled models, but steering was difficult. The Lindeman Model D was not put into production, but the partnership with John Deere lasted for decades. *Lindeman Archives*

MODEL DX24 EXPERIMENTAL CRAWLER TRACKS

This machine was built as a crawler in 1930, and shipped to Havre, Montana. The machine was rebuilt with wheels on April 20, 1932. The track carriers were fabricated by restorers Justin and John Kutka using period photographs. The track chain and pads are sourced from a 1950s John Deere crawler.

The D's horsepower had been improved a bit in 1928 with the addition of a quarter-inch increase in the cylinder bore and an improved carburetor, resulting in an increase in the drawbar horsepower to about twenty-eight, and thirty-six horsepower on the belt. Unfortunately, Wiman reported, "No more horsepower could be built into our present Model D tractor without a pretty thorough redesign throughout . . . this tractor is now up to its limit of strength and stability with the horsepower delivered to the drawbar as at present. . . . If the IH is able to beat us in the field-on-field performance . . . we will have to rely on our selling of these machines in volume on the greater simplicity of the Model D tractors, its two-cylinder construction, and its reputation as a low-cost machine from an economy standpoint and from the non-use of a large number of repairs."

INCREASED POWER MODEL D

The Model D engine received a power boost during production. *Lee Klancher*

The horsepower fears were exacerbated when in January 1929 Deere learned that International Harvester had also incorporated a quarter-inch increase in the bore of the cylinder of its 15-30 tractor, which it was rumored would add anywhere from five to twelve horsepower to the machine. This worried Deere not only because of the disappointing horsepower of the GP, but also as a direct assault on the Model D. Inasmuch as the D was the large-production star of the Deere line, any threat to its sales was even more serious. Wiman reported to the board, "It is unfortunate that this vicious circle or race for horsepower has again been started, but it is not strange when you consider the fact that for the last two years we have been able to outperform the International 15-30 machine, that they should increase the speed and power on their machines to equal or better our own. To sell their old 15-30s on the territory, the International Company have cut the price of this machine one hundred dollars and we believe this will have to be cut further yet."

With the renewed threat from the 15-30's increased horsepower, Silloway decided to reemphasize the competitive advantages of the Model D and issued a bulletin headed, "John Deere, a two-cylinder tractor." The memorandum bluntly denied that Deere was thinking of shifting to a four-cylinder tractor. "Why should we?" Silloway wrote, "The John Deere two-cylinder tractors will do plowing and other field work at the lowest possible cost per acre and, after all, that is what the farmer is interested in." Silloway emphasized the light weight that led to fuel economy, the extreme simplicity, with its low cost of upkeep, and the fact that the owner could do his own servicing.

Detractors of the two-cylinder concept often used as their argument the excessive vibration of a two-cylinder motor. But Deere counteracted this attack by sending a Model D tractor to fairs, mounting the machine on four pop bottles, and putting the tractor in operation with the rear wheels turning. There was not enough rhythmic vibration to shake the machine off the mouths of the bottles.

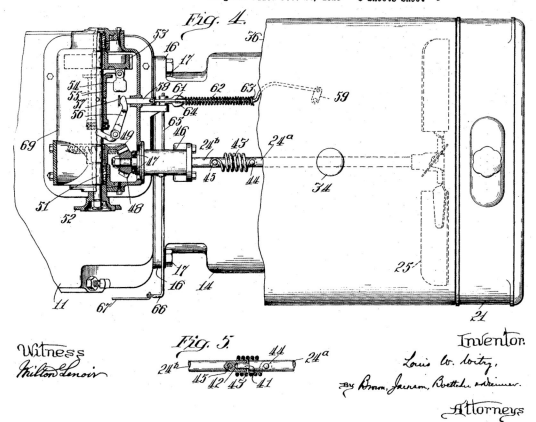

March 4, 1930.

L. W. WITRY　　　1,749,202

TRACTOR

Original Filed Dec. 24, 1923　　3 Sheets—Sheet　3

Fig. 4.

Fig. 5.

Witness
Milton Lenoir

Inventor.
Louis W. Witry,
By Brown, Jackson, Boettcher & Dienner.
Attorneys

STYLED MODEL D

The radiator, dash, and fenders of the Model were redesigned for the 1939 model year, and a new drawbar and platform was added.

John Deere Archives

> *"There is a national demand for tractors. We do not have to create it, and when a suitable tractor is built at a reasonable price to the consumer it can be sold."*
> —Leon Clausen, Deere executive, about the Model D, April 1922 board meeting

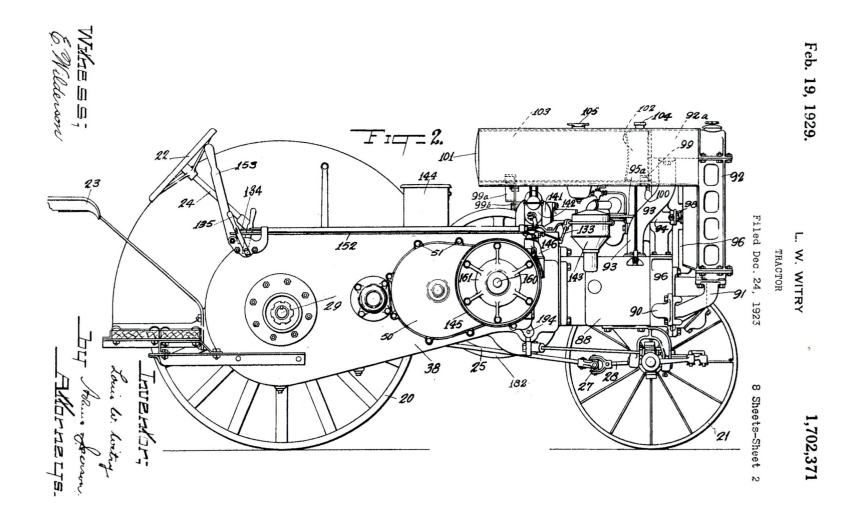

MODEL D DATA

Model	Type	Model Years	Notes	HP	Nebraska Test #
D (Spoker)	Row-Crop	1923–1925	First production 05/30/1923. Last production 12/28/1925. Kerosene engine (gas start).	30	102
Unstyled D	Row-Crop	1925–1939		30	146
XD Experimental	Row-Crop	1928	100 built. First built 08/17/1928, last known built 10/19/1928.	-	-
DI	Industrial	1936–1941		30	-
Styled D	Row-Crop	1939–1953	First production 04/07/1939. Gasoline engine kit offered 1951 and later. Last one built 07/03/1953.	30	350

CHAPTER TWO

THE
FOUNDATION

LETTER SERIES TRACTORS

*"Cheaper power is one of the greatest needs of
agriculture to-day . . . the general-purpose tractor
will open the gateway to cheaper power."*
—"The Design of a General-Purpose Tractor",
H. B. Josephson, Graduate Thesis,
Iowa State College, 1925

1935 MODEL BN

Dave Nelson Collection / Lee Klancher

The tractor would come of age in the late 1920s and 1930s, with the machine transitioning from an expensive specialty item for co-ops, massive operations, and the rich to an essential farm tool.

Battling the epoch of IH's tractor technology development, the challenges brought on by the first World War, and the brutal economy of America's Great Depression, Deere would transition its caution in developing the farm tractor into positioning itself as the manufacturer who came closest to the massive and growing conglomerate that was IH.

The transition was a remarkable one. Deere was famously conservative about entering the tractor market, and many had taken it to task for waiting so long to build its own. History has painted CEO William Butterworth, in particular, as a detractor of the machines. Historian Wayne G. Broehl Jr. quotes

a handwritten note of Butterworth's in which he is writing about his strategy to deal with 1912 business difficulties. His last line of that document stated, "Drop all tractor expenditures."

At that time, his skepticism proved prescient. The Deere experimental tractors of the time were not promising, and the company would find itself with a mountain of inventory of implements shortly after. That was most certainly not the right time to invest in new, unproven technologies.

Only a few years after Butterworth smartly backed down, General Motors (GM) let pride lead it to attempt to follow in Ford's footsteps and build a tractor. It merged several companies to form Samson Tractor in 1918, and proceeded to pour money into tractor innovation. The results included a belt-driven, four-wheel-drive model powered by a Chevy 490 that was steered

MODEL GP SKETCH

Theo Brown sketch of the GP tractor. *Theo Brown Archives / WPI*

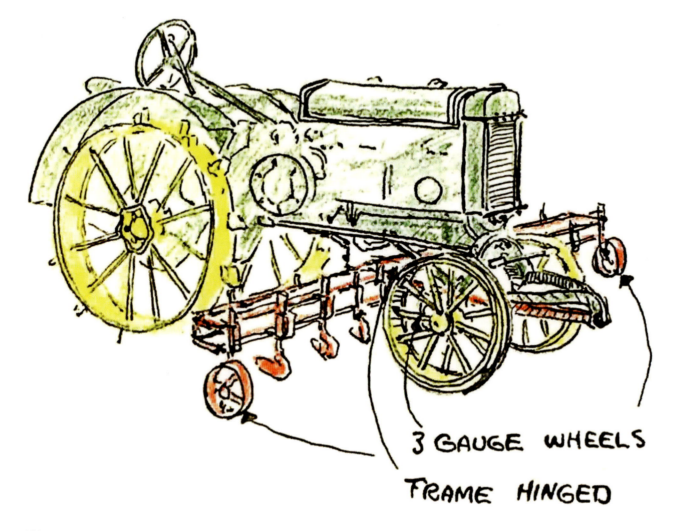

3 GAUGE WHEELS

FRAME HINGED

with a pair of reins. The venture was an abject failure, and GM reportedly racked up $33 million in losses on its tractor exploration.

Butterworth was steadfast as Dain's tractor developed in 1918, writing in a letter to colleague Burton Peek, "I am opposed to any step being taken towards the manufacture . . . of a tractor." His reasoning was based on their market position. Given Ford's deep pockets, he did not want their limited capital dedicated to "an awfully expensive enterprise which will get us nowhere in the end."

Many in Deere, including Peek, felt this was a mistake.

While it is certainly true that tractors would become a very big deal, competing with the behemoth that was IH and Henry Ford's willingness to empty his pockets to dominate the tractor market, perhaps Butterworth was more prudent than history paints him.

In the end, the Waterloo Boy, the Model D, and a worldwide wave of power farming development won over Butterworth. He would stay on long enough to see a dizzying transition, as Deere would uncharacteristically scramble to match a competitor's brilliant innovation and follow that up with a series of tractors that would power farmers around the globe and become one of the most prized collectible machines in history.

MODEL GP

In the late 1920s, all the major manufacturers scrambled to build a general-purpose farm tractor. *John Deere Archives*

MODEL GP

The Model D was a strong success, but Deere & Company's cautious approach to farm tractor production had cost it. The loud, crude Fordson had put farm tractors in the minds and into the landscape of the average farmer, convincing most of them that the high technology was attainable and, for some tasks, superior to horses and mules.

In 1924, International's brand-new general-purpose Farmall delivered on the Fordson's promise with a machine that could turn a belt, pull a plow, and cultivate. The tall back wheels and good visibility gave it the ability to cultivate crops. This simple change revolutionized the industry, and once customers got over the tractor's unfamiliar appearance, the Farmall pushed the Fordson out of the market entirely.

The Farmall's sales figures had the entire industry scrambling to build similar general-purpose tractors.

With Butterworth in a less active role and Deere having a firmer grasp on the number two position in the industry, Charles Wiman assigned his brilliant engineer Theo Brown the task of building a general-purpose machine.

Brown was one of the company's star engineers. He had led the experimental plow division in the teens, and he eventually was called on to design many of Deere's most innovative machines and implements. Brown was one of the leading lights of tractor design in the early twentieth century, and his brilliance shines particularly bright in modern times due to his diary. Each day, he recorded his thoughts, clipped photographs, and drew sketches of his ideas and impressions. This fantastic record of tractor innovation lives on as the Theo Brown Diaries housed in the archives of the Worcester Polytechnic Institute, and the diary provides an intimate look at the development at John Deere.

1927 PROTOTYPE

This photograph of a Model C tractor with a three-row planter is dated May 1927.

John Deere Archives

1927 MODEL C

The Model C was the forerunner to the Model GP, and production began with seventy-six built in 1927. Another group of Model Cs were built in 1928, many of which were recalled 1927 models that were rebuilt in the 1928 run. The last Model C was built on April 20, 1928. The Model then was renamed the Model GP. In 1931, Deere declared Model Cs "experimental." The machines were supposed to be sent to the company and rebuilt, but at least one (seen here) survived in original condition. This example was serial number 200193, a production Model C. *Keller Collection / Lee Klancher*

EARLY CHALLENGES

The Model C was developed quickly and had significant shortcomings. One of the features was an unconventional three-row cultivator. Most cultivators available at the time were designed to work two or four rows, and the three-row model was not widely accepted or well regarded. The Model C engine was a twin-cylinder of ill repute. Power output was deemed inadequate.

The tractor that Brown developed—first the Model C experimental, which became the Model GP—also is arguably the most interesting model in the line. Without a doubt, it has the most convoluted production history of any single model, at least in its relatively short lifespan, and changed continuously throughout its life.

Brown began the task of building the general-purpose machine on September 30, 1925. He traveled from Waterloo to Ames, Iowa, to meet with other Deere & Company design leaders. They reiterated the market need for a cultivating tractor, and analyzed the failure of the Moline Universal to establish parameters for building a successful machine. Brown adopted the novel concept of creating a three-row cultivator, an act that came to be considered a mistake.

On October 5, Brown wrote that he and his team believed the new tractor could use a wide-front design with a cultivator that bolted to the frame, just ahead of the front wheels. By October 9, he was sketching his idea on paper. On the sixteenth, Brown built a model of the proposed "All-Crop" tractor and was pleased with the results. A few days later, the team fastened a crossbar rigged with cultivators to the front of a Fordson. The hand-built unit tested better than expected.

Brown spent December 1 at the factory inspecting a full-scale mockup with Deere & Company President Charles Wiman. "McCray has built a model 'All-Crop' tractor of wood which is complete as to looks—painted yellow and green," Brown wrote. "The outfit looks good and the center of gravity not so high as we feared. The top of the radiator stands three inches higher than the Model D. In two weeks they are to send the model over here and we will put on the cultivator. If everything turns out well we hope to have a tractor in the field by May 1st."

After several months of furiously revising sketches of the All-Crop, Brown spent a full day on May 19, 1926, testing the new prototype. Refinement was required and ensued steadily. By August 14, Brown wrote, "Our present design is about right."

As the year wore on, Brown recorded a string of improvements along with demonstrations of his machines and all-day meetings debating the merits of various improvements and changes. The three-row cultivator was constantly debated, but never rejected.

In the spring of 1927, testing and comparison continued. Linkages were refined, and the radiator was widened.

Dealers clamored for the new tractor and cultivator, and the pair was listed without prices in trade catalogs. An August 1927 release informed dealers that experimental models were being tested but were not ready.

On December 15, Brown wrote that the "first Model C was run off today." On February 23, 1928, his dissatisfaction with the machine was evident in

his journal. "It seems to me that the view is the real problem now," he wrote. A few days later, after testing a tricycle front-end model, he felt there was progress on the principal problem. "The dodge is fast and the view is good," he wrote on February 29, but he wasn't satisfied, nor was the problem resolved. Brown never would find a satisfactory answer to the problems on this machine. In a move uncharacteristic for Deere & Company, the machine was sent out into production.

The first production Model C appeared on March 15 of that year. Its production span was incredibly short-lived, and came to an end on April 20. A few months later, Deere & Company issued a notice to its dealers that the name of the Model C was being changed to Model GP (for "general purpose").

Frank Silloway's memo about the change spells out the fact that the Model C moniker was being dropped due to complaints from dealers that C and D sounded too similar and taking orders over the phone could be confusing. He also noted that they had considered and rejected naming the tractors by horsepower ratings or by a clever name.

"We do not care to popularize with a name such as 'Powerfarmer' the smaller tractor on which we make little profit, as against the Model D, which is the profitable tractor, for both the dealer and ourselves to

MODEL GP FIRST PRODUCTION

The production model of the Model C was dubbed the GP. This is another extremely rare model, the first Model GP built. It is serial number 200211.

Keller Collection / Lee Klancher

sell and the one that most farmers should buy. The chances are the Model D will always be the one which will make our tractor endeavor worthwhile," Silloway wrote.

"If we do give any new tractor some distinctive name, it must necessarily be a name which describes its more general utility in farm operations, rather than a name which will in any way detract from our principal tractor and the one in which our interest will always be centered, namely: the Model D."

The interesting reading between the lines here was that Silloway and the Deere management team were not entirely sold on the GP, or at least did not believe the sales potential exceeded the D. While this seems foolish now, bear in mind hindsight is twenty-twenty. Also, the Farmall was making waves by 1928, but it did not become widely adopted as the next Big Thing (even by IH management) until 1930.

The key point here is tractors were a rapidly changing new technology in 1928. Visionaries and college students recognized what was on the horizon, but that wasn't the broad population. And the kind of men selected to lead very large companies, with large sums of money at risk and huge groups of shareholders concerned about returns, are prone to be level-headed sorts unlikely to make rash decisions.

Silloway was not overly cautious—he in fact was the one who pushed for the purchase of the Waterloo Boy Company—and was clearly more progressive than the staid leader Butterworth. Silloway's words are merely a reminder that general-purpose tractors in 1928 weren't so different than personal computers with a graphic user interface in 1984. Both were radical ideas that would change their industries . . . and both were received with widely shared doses of skepticism.

MODEL GP

The Model GP saw significant evolution during production, as the company constantly made changes to improve the machine.

MODEL GP WIDE-TREAD

By Wayne G. Broehl Jr., from *John Deere's Company*

Deere executive Charles Wiman's skepticism about the GP was confirmed by reactions from farmers. To start with, there were a number of breakdowns in operation, requiring rebuilding of some of the tractors in the field. Even more important was the farmer reaction to the three-row cultivating feature. While many farmers in the corn belt seemed to accept the notion reasonably well, the cotton growers in the South clearly preferred two- and four-row operations. There were also inherent design problems; one was visibility for the driver on the GP. "It seems to me that the view is the real problem now. The Farmall has the best of us there," noted Theo Brown. Wayne H. Worthington, in his definitive study of the agricultural tractor for the Society of Automotive Engineers, succinctly summed up the reputation of the GP: "Unfortunately, the three-row idea failed in its acceptance and the tractor fell far short of Model D performance and durability."

Within a year, Brown and his development team were back at the drawing boards trying to rectify the GP's problems. A crash program soon brought into production a variation of the GP—the GP Wide-Tread. In many respects it was similar to its predecessor except that it had a long rear axle that allowed the machine to straddle two rows and the first John Deere tricycle front to run between two rows, just as the Farmall did. By the crop season of 1929, the GP Wide-Tread was available to buy, and its acceptance not only in the South but in the Midwest was more gratifying. In 1931 the GP itself was modified to raise its horsepower ratings to 15.52 drawbar, 24.30 belt-pulley. A year later, in 1932, the GP Wide-Tread was modified to provide a tapered hood to increase the visibility of the driver and to provide changes in the steering apparatus that prevented the tendency of the front wheels to whip on rough ground. With the second model of the GP, and the modifications and additional features, the GP finally made a respectable showing in the Deere line. Charles Wiman always remembered his experience with the GP, however, as "an outstanding failure" and often mentioned: "Well do I recall how much tractor business was lost by our company due to bad design of the 'GP' line."

MODEL GP-WT SKETCH

The key weakness of the GP was it was set up to be a three-row machine. In the late 1920s, Deere assigned engineer Theo Brown to address the issue. This April 30, 1929, sketch shows Brown's early efforts on the new four-row model. On May 1, he wrote, "Wiman dislikes to appropriate more money for equipment for wide tread tractors and power lifts where there is no profit." Charles Wiman, a Deere executive, was mainly right about the WT's profits, which would be thin, but the power lift would pay off handsomely in royalties due to winning a patent infringement case against IH (see sidebar, "Theo Brown's Power Lift", page 76). *Theo Brown Archives / WPI*

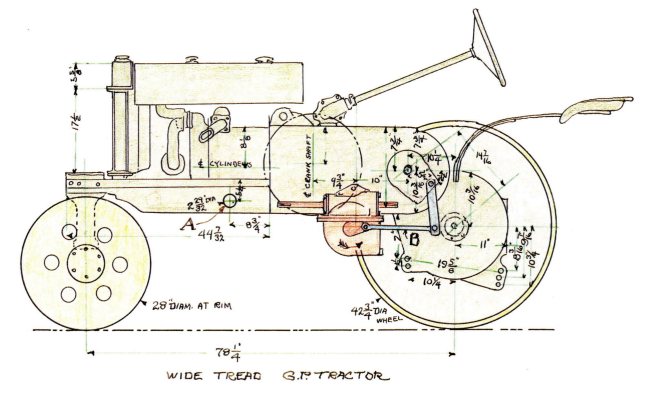

WIDE TREAD G.P. TRACTOR

MODEL GP-WT EXPERIMENTAL

The distance between the rear tires needs to vary for different crops, yet early tractors had fixed axles that didn't allow any variation. The logical solution was to allow the wheels to slide on the rear axle. This experimental model looked to solve that with a power-adjustable rear axle. The model was never produced, and this machine is the lone survivor from that experiment. *Keller Collection / Lee Klancher*

MODEL P

One of the rare variants of the Model GP-WT is the Model P (or Model GP-P), which was a GP-WT modified for potato farming with a narrower 68-inch rear tread width, different wheels and axles, and a few other bits. According to *Two-Cylinder* magazine, 203 of these were built in 1930. This is the second production Model P, serial number 5001.

Keller Collection / Lee Klancher

1930 MODEL GPO EXPERIMENTAL

The orchard version of the Model GP is a rare machine, and this example has a mysterious history. The Deere records show that, in 1930, six experimental GPOs were created by converting GP-WT tractors, and those were issued serial numbers 14994 to 14999. Regular production of the machines proceeded from 1931 to 1935, and roughly 718 regular production Model GPOs were built and assigned serial numbers from 15000 to 15732. The Deere records indicate regular production serial numbers 15220 to 15225 were not used. Here's the interesting part: the machine shown here has parts that clearly indicate it is one of six experimental GP-WTs pulled from inventory in 1930 and converted to orchard models. The smoking guns are the long elbow on the air intake and a relocated oil drain plug on the final drive case, both features being unique to these six experimental models. The serial number on this tractor is 15223—one of the "missing" regular Model GPO production numbers. There is another experimental GPO in Oregon that also bears a serial number from the "missing" number sequence.

Keller Collection / Lee Klancher

1933 MODEL GP-WT

In 1929, the GP-WT was introduced. The model still lacked horsepower and had poor visibility, which led to more modifications as the years went on.

Mauck Collection / Lee Klancher

HARD LESSONS

More than thirty-five thousand units of the Model GP sold between 1928 and 1935, and the experience taught a segmented company the importance of developing tractors and implements in concert. The machine also gave John Deere a much-needed general-purpose tractor. The Model GP had serious flaws—the three-row cultivator was the wrong configuration from day one and the engine was underpowered. Some of those issues were addressed with nearly continuous development. Even with the improvements, the GP never quite matched Deere's high standards. Wiman expected—and Brown would find—more satisfactory results with future machines.

Function and sales were the goals of Brown's three-year odyssey of sketches, models, and testing. His labor resulted in a machine that offered Deere & Company hard-earned lessons about the new craft of large-scale tractor design.

Viewed in the proper light, the imperfect but beautiful Model GP is a rolling work of art, and an expensive and troublesome reminder that building industry-leading equipment requires long development cycles. This is a lesson that John Deere leadership and the engineering teams would take to heart for decades to come.

MODEL GP-WT
TOP STEER

Another production change to the GP—and there are so many it's hard to track—was to move the steering on top of the hood. *Brad Bowling*

MODEL GPO LINDEMAN CRAWLER

The Lindeman brothers of Yakima, Washington, were contracted by Deere to convert about twenty-five GPO models with tracks. The Lindemans soon switched their conversion efforts to that machine. The GPO Lindeman Crawler is a very rare machine today, with about a dozen known to exist.

MODEL GP DATA					
Model	Type	Model Years	Notes	HP	Nebraska Test #
C Experimental (Early GP)	Row-Crop	1927–1928	C renamed Model GP 06/28/1928. Kerosene engine.	22	-
GP Standard	Row-Crop	1928–1935	Engine bore increased from 5.75 to 6 inches (more power) on machines built after 08/1930. Kerosene engine.	25	153 (10-20), 190
P (or GP-P)		1930	203 built, converted Model GPs. Kerosene engine.	25	-
GP Wide Tread		1929–1933	Kerosene engine.	25	-
GP Lindeman Crawler	Orchard	1931–1935	Kerosene engine.	25	-
GPO (Orchard)	Orchard	1931–1935	24 GPO converted into Lindeman Crawlers	25	-

MODEL GP-WT

The Model GP and variants were a rare example of John Deere rushing a machine into production, with predictable results. Despite their flaws, the machines served the company and thousands of farmers well. Perhaps more significantly, they provided an early lesson in the importance of Deere sticking to its long-held belief that developing good products requires long development cycles. *Mauck Collection / Lee Klancher*

This piece provides superb context on the tractor market in the 1920s, and illustrates the fascinating point that Deere has battled red tractors (IH and later Case IH) for the top two market positions in agricultural equipment for nearly a century. What other industry has been dominated by so few for so long?

The decade of the 1920s was a watershed for the agricultural machinery industry. First, there were major consolidations. In 1928 the Canadian-based Massey-Harris Company took over the J. I. Case Plow Works in Racine, Wisconsin; Massey immediately sold its rights to the "Case" name to the J. I. Case Threshing Machine Company, also in Racine, and kept for itself similar rights to the prestigious Wallis tractor. In turn, the J. I. Case Threshing Machine Company bought the farm implement business of the Emerson-Brantingham Corporation—the old-line Rockford, Illinois, implement manufacturer that had been in continuous operation since 1852—and then changed the combined company's corporate name to J. I. Case Company, Inc. The next year, 1929, saw another new company organized—the Oliver Farm Equipment Company, a merger of the Oliver Chilled Plow Works of South Bend, Indiana; the Hart-Parr Company of Charles City, Iowa; and the American Seeding Machinery Company and the Nichols & Shepard Company of Battle Creek, Michigan. Oliver, continuing to make its own chilled plow (of which it was a pioneer), now added grain separators and farm tractors. In this same year, yet another new long-line company came into being, the Minneapolis-Moline Power Implement Company—a consolidation of the Moline Implement Company, the Minneapolis Steel and Machinery Company (builder of the Twin City tractor, the machine that briefly had been sold by Deere in the 1910s), and the Minneapolis Threshing Machine Company (also a builder of tractors). In this same period, Allis-Chalmers Manufacturing Company of Milwaukee, Wisconsin, added to its tractor line by purchasing the business of Monarch Tractors Corporation of Springfield, Illinois, manufacturer of a track-laying tractor, and then becoming a full-line manufacturer with the acquisition in 1931 of the Advance-Rumeley Thresher Co. of LaPorte, Indiana.

In 1921, there had been 186 tractor companies; by 1930, the number had been reduced to 38. Whereas there had been many dozens of tractor companies and many hundreds of implement companies in earlier years, now there were just seven main companies offering full lines— the "long-line" companies were Deere, International Harvester, Case, Oliver, Allis-Chalmers, Minneapolis-Moline, and Massey-Harris.

A second stage of realignment came when every one of these newly combined companies embarked on the quest for its own general-purpose tractor: in 1930 Oliver introduced its new tricycle Row Crop, Massey-Harris came out with a four-wheel-drive general-purpose model, and Allis-Chalmers announced its All-Crop general-purpose tractor. The playing field of tractor competition was fast becoming crowded!

Deere, meanwhile, had not only bought two small companies (the potato harvesting equipment of the Hoover Manufacturing Company and a threshing line, purchased from the Wagner-Langemo Company), but had added more than $7.6 million to existing capacity in its factories and branches. By 1930, the company had opened new branches in Fort Wayne, Indiana; Amarillo, Texas; and Edmonton, Alberta, and had also constructed warehouses in Lansing, Michigan; Sidney, Nebraska; Aurora, Illinois; and Swift Current, Saskatchewan—quite a change from Charles Deere's earlier marketing philosophy of a few large branch houses.

The 1920s witnessed major shifts in market position. The long-line companies picked up increasing percentages of the market for most of the key products. Deere and International Harvester were particularly strong; together they dominated most product categories, with Deere moving up faster in many of these. A Federal Trade Commission report made these public.

The reasons for dominance by the larger, long-line companies, in the opinion of both industry and outside analysts, lay particularly in the strong marketing structures of these leaders, with the branch house as the linchpin. Warren Shearer's comprehensive study of the industry puts particular emphasis on the 1919–29 period as "the formative decade." Shearer broke down the extensive documentation in the FTC report (cited above) to show that there were indeed sharp differences among the long-line manufacturers in regard to both manufacturing costs and marketing costs, with the latter particularly striking. He used 1929 for comparative purposes, as shown in Chart 2.1.

PULLING STRONG

In 1921, 186 makers were building tractors and Deere was just getting started producing the machines. By 1929, John Deere was an industry leader. By the twenty-first century, Deere was the only company to survive without being purchased, merged, or dissolved. *Library of Congress, LC-USF33-012101-M3*

A GROWING LINE

After being cautious about creating new models of machines in the 1910s, Deere steadily grew its tractor line through the 1920s. *John Deere Archives*

TOP THREE TRACTOR BUILDERS MARKET SHARE 1916-1929

Year	Total Tractor Production	John Deere Market Share	International Harvester Market Share	Ford Market Share	Total Market Share Controlled by Deere, Ford and IH
1916	29,670	0.0*	39.0**	0.0	39.0
1918	132,697	4.2*	19.0**	25.6**	48.8
1921	68,029	0.0*	26.1**	19.1**	45.2
1929	223,081	21.1	59.9	0.0	81.0

Source: Federal Trade Commission. Report on the Agricultural Machinery Industry. Portions of the chart are from *John Deere's Company* by Wayne G. Broehl Jr.

*Deere produced 5,634 Waterloo Boys in 1918 and 79 Waterloo Boy Tractors in 1921; market share calculated using those numbers and total production.

**1916, 1918, and 1921 Ford and IH tractor market share calculated from IH production numbers and industry-wide numbers from *The Agricultural Tractor* by R.B. Grey.

FROM ZERO TO HERO

This chart shows shows how John Deere grew from no tractor market presence in 1916 to become the second largest tractor maker in 1929. This was an incredible feat in a crowded, competitive market and a time of frenzied growth in tractor companies. To make the situation more difficult, the top two competitors were a massive conglomerate (IH) and a megalomaniac billionaire more interested in making a point than generating profit (Ford). As the volume of tractor sales grew during the 1920s, the brutal competition thinned the ranks of makers. By 1929, only 47 tractor companies survived, the tractor market grew tenfold in unit sales, and Ford was chased out, with Deere and IH the dominant players.

Shearer's study confirmed the striking strengths of Deere and International Harvester in the 1920s. He concluded, "It has never been charged that IH and Deere had power to restrict production or set limits on the output of their competitors. It is equally true that their pricing policies were not of a sort to discourage competition except in the sense that such competition was unable to meet Deere and IH prices and still make a profit. If we are to discover any control over the market exercised by these two acknowledged leaders it must be found in the distribution system which characterized the industry. . . . If we neglect the possibility that these two concerns produced a markedly superior product, and it is fair to assume that this was not the case, the great strength of IH and Deere must have been their dealer organizations or the control they exercised over these organizations."

There were technological product breakthroughs for certain products, for example, the general-purpose tractor of International Harvester. Still, Shearer's judgment corroborates the findings of a number of other analysts: it was the large, strong, long-line companies that were able to develop comprehensive distribution systems and use the branch-house concept to its fullest, and thus to develop a marketing relationship with both dealers and farmers that gave them greater strength than the smaller long-line companies. Most of the short-line companies were not able to mount extensive dealer networks that included branches (though some were able to maintain their market share by virtue of patent protection).

The strikingly low selling costs of the three industry leaders (Deere, International Harvester, and Allis-Chalmers) testify to the substantial economies of scale in the larger branch-house systems. Yet, these three were also serving their dealers, and the latter serving their customers, the farmers, in a more effective way than the other long-line companies. Deere, in particular, never forgot the lesson that John Deere had learned so well—it was the farmer who would be the ultimate arbiter of success for the company.

"With everything so quiet it is very hard to get any enthusiasm in work. It is depressing to learn so much about hard times, cutting expenses, etc. An experimental man needs enthusiasm to do good work. And I can see how this depression affects original thinking."

Theo Brown, writing downheartedly in his diary in 1931, captured the essence of the dilemma facing a researcher in those darkest days of the Great Depression. How can one feel excited about new ideas when everything around him cries, "Caution! Cut back! Don't do!" Susceptible to depression in this period, Brown confided to his diary his frustration and sense of a loss of purpose.

Several longstanding company experimental endeavors were indeed severely curtailed, including the soil culture department and several experimental farms in the Moline area. The former had been the brainchild of Dr. W. E. Taylor, who had started it for Deere in 1910 and had developed a set of company publications on soil culture that were advanced concepts in the field. Taylor, a prolific speaker, built a strong following among soil scientists around the country. By 1931, he was ready to retire, and it seemed appropriate, given the economic situation, to close the department. Work in soil science continued in the advertising department, though at a slower pace in the 1930s.

But Charles Wiman had no such doubts about basic product research. Of all Wiman's management decisions during the Great Depression, the one that was probably most long lasting in effect was his decision to aggressively continue new product development through the worst days of the downturn. Already the Model D tractor was highly popular with the trade—the company had sold more than 100,000 units by 1930 and would keep the model in the product line twenty-three more years after this date. The GP and its companion, GP-Wide Track, had not done as well, though more than 35,000 would eventually be made before the latter was terminated in 1933, the former in 1935. Still, competition from the other companies constrained one from resting on laurels with one good model and two somewhat limping models. Also, it was becoming increasingly apparent that farmers around the country were avidly looking for many improvements—adjustable tread widths in the rear wheels, less side draft in the tillage instruments, perhaps even smaller versions of existing tractors.

By 1931, Wiman had put Brown and the other engineers to work on new tractor ideas, and a few months later, in April 1932, he sent Brown on a special trip to Dain, Van Brunt, Syracuse, and the Wagon Works to "pep them up somewhat on experimental work." Out of Wiman's constant press for experimental and engineering effort, and his exhortations for fresh ideas, came two new tractors, the Model A (16.22 drawbar horsepower, 23.5 belt-pulley horsepower) and the Model B (9.28 drawbar horsepower, 14.25 belt horsepower). Tested in the Salt River Valley in Arizona in 1933, the Model A was brought into production in the following year; one year later the Model B was introduced.

The two tractors were strikingly successful; both would stay in the product line until 1952, with more than 293,000 of the As sold by that date, more than 309,000 of the Bs. The two tractors rank first and second in popularity over the entire tractor history of the company.

HARD TIMES

Deere wisely invested in research and development of tractors during the Great Depression. When the hardship began to recede, Deere was ready with new equipment.
Russell Lee, Library of Congress LC-USF33-012358-M4

MODEL A

By Wayne G. Broehl Jr., from *John Deere's Company*

The Model A was the solution to Deere's need for a two-plow tractor. Its adjustable wheel tread answered the farmer's need for moving his wheels outward or inward from the then-typical 42-inch standard row. (There had to be this much room for the farm horse to continue to walk on solid ground for each pass through the field.) The implement hitch for the A, as well as the power shaft, was located on the center line of the tractor, substantially reducing any side draft. An industry first, a hydraulic-power lift system increased both the efficiency and speed of operation, as well as providing a "cushion" drop of equipment. There were other important features, for example, a one-piece transmission case that allowed high under-axle clearance.

The simple, powerful two-cylinder engine successfully burned distillate, fuel oil, furnace oil, and similar low-cost fuels, as well as kerosene or gasoline. Thus the engine could be started with gas, then switched over to the lower grade fuels. These cheaper substitutes burned more effectively in the big cylinders of the two-cylinder engines than when used in the conventional four- and six-cylinder engines. This was a period of hard times, and an economical design with economical fuel was particularly important. Wayne Worthington, in his definitive study of the tractor, evaluated Deere's decision: "Deere & Company had always taken a contrary position with respect to fuels, and as others turned to the use of Regular gasoline (70 octane) they continued to promote the economies made possible by the use of available low cost distillates. A continuing program of combustion research was followed, which resulted in increasing the compression ratio of their distillate burning engines by some 40 percent. The resulting fuel economy broke existing records when tested at Nebraska. This served to educate tractor users to the importance of fuel economy even though the price of tractor distillate delivered to the farms was in the order of eight to eight and a half cents per gallon."

JOHN DEERE GX EXPERIMENTAL SKETCH

Theo Brown presented proposed specifications for a new model at a February 1931 Deere board meeting. Brown believed the machine could not meet goals without a four-cylinder engine, but the board deemed a new engine too costly. Deere's Elmer McCormick began to develop a new tractor coded the GX, which eventually became the Model A. This sketch is the proposed design by Theo Brown, in his June 25, 1932, diary entry. By September 29, 1932, the machine was being tested in the field, prompting Brown to write in his diary, "The GX is so far ahead of 32 W.T. [GP-WT] that there is no comparison."
Theo Brown Diary / WPI

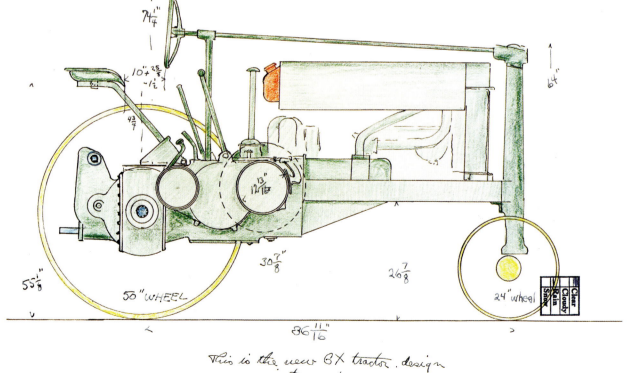

1937 MODEL A WITH CORN PICKER

Deere had a strong showing in the corn picker market, owning 29 percent of the market in 1929. It was even stronger in corn planters, holding 41 percent of the market—more than agricultural giant IH. This fascinating early Model A was photographed at the Half Century of Progress Show in August 2019.
Murrell Collection /
Lee Klancher

1937 MODEL A WITH
CORN PICKER

FIRST PRODUCTION MODEL

This Model AR is serial number 250000, the first production model built. The new standard-tread Model A was advertised in a 1935 brochure showing a leaping deer logo on the hood between "John" and "Deere." According to an article in *Two-Cylinder* magazine, none of the Model ARs were produced with the leaping deer logo.

1937 MODEL AWH

One of the rare Deere models is the AWH, with only twenty-seven of them built at the Deere factory. The AWH has an adjustable tread width. The machine's narrow width, combined with a wide tread width, make the model ideal for cultivating vegetable crops.

Keller Collection / Lee Klancher

MODEL AWH

The Model A uses a two-cylinder 309-ci all-fuel engine that starts on gasoline and can run on cheaper fuels. The carburetor on this AWH is a Marvel-Schebler DLTX-18. Power is delivered with a four-speed transmission that yields a top speed of 6.25 miles per hour. A Model A weighs in at about 4,000 pounds and in 1937 cost $1,050—not cheap at the time.

MODEL AOS

The Model AO Streamlined (AOS) was the last gasp of Deere's engineer designers. Created under the watchful eye of chief engineer Elmer McCormick, the machine is an oddly appealing juxtaposition of curvaceous sheet metal on top of a rather stolid Model AR.

To understand the model, it's best to take a step back and understand the AO, which is a mildly complicated beast. The first AO was created hand-in-hand with the AR, with the first of those machines built in spring 1935. This early unstyled Model AO had a lower air intake and a muffler that vented to the side rather than above the hood as well as different brakes. The first of these was built on May 22, 1935. Production of this variant lasted only until October 1936, when the Model AOS replaced the Model AO.

1937 MODEL AOS

In 1936, six experimental Model AOS tractors were built, tested, and returned to the John Deere Tractor Works in Waterloo in 1937. Two were scrapped, while the other four were rebuilt as production models and sold. The 1937 Model AOS pictured here is a regular production model. *Keller Collection / Lee Klancher*

NARROW ENGINEERING

The Model AOS is narrow for orchard work. That narrow profile necessitated changes to the clutch assembly, rear-axle housing, crankshaft, and front axle. In addition, the crankshaft was redesigned with a new flywheel and other parts, helping to shave off another half inch of width. The right main bearing was redesigned, and the crankshaft bearings were rotated 180 degrees.

The Model AOS was significantly modified for orchard use. It was six inches narrower, five inches lower, and the wheelbase was shortened by seven inches. These have their own serial numbers, and some have fenders that cover the top half of the wheel, while others have the small, curved fenders on the model shown on these pages.

The jaunty small fenders add a sporting flair to the machine, which led artist Charles Freitag to create a painting of an AOS on a banked racetrack, battling

REAR VIEW

To narrow the Model AOS, the axle shafts and housing from the Model AR were shortened. The drawbar assembly was also redesigned. The lever for shifting into low gear is bent to provide clearance under the lowered and shortened steering column. The seat is larger and lower than the unit found on the Model AR Standard.

with period open-wheel cars, the pilot hammering down the turn wearing a pith helmet as his scarf trails behind him in the breeze.

In 1940, the Model AR received an update, with the most notable feature a more powerful engine. Deere dropped the Model AOS and replaced it with a Model AO that looked very similar to the 1935–1936 machine.

Finally in 1949, Deere released the styled Model AO, which was designed by Henry Dreyfuss Associates (HDA). The look is a clearly delineated line that runs directly counter to the philosophy of Industrial

Designer Henry Dreyfuss, who would begin work with Deere & Company in August 1937, long after the AOS was off the drawing boards and into the fields as a test mule. Dreyfuss believed form must follow function, and his design contributions to John Deere would be simple and elegant

The styled Model AO is considerably more refined and elegant than the gaudy AOS. Flashy fenders and droop-nose grilles are better suited to Buicks and Buck Rogers films than tractors.

MODEL A DATA

Model	Type	Model Years	Notes	HP	Nebraska Test #
AA	Row-Crop	1933	Experimental. Only 8 built. First AA built 04/15/1933.	NA	-
Unstyled A	Row-Crop	1934–1938	First production (gas) 07/14/1934; last production (gas) 05/12/1952. All-fuel engine.	24	222 (1934)
Unstyled AN	Row-Crop	1934–1939	Only 591 built.	24	-
Unstyled AW	Row-Crop	1934–1939	Only 303 built.	24	-
Unstyled ANH	Row-Crop	1934–1939	Only 26 built.	24	-
Unstyled AWH	Hi-Crop	1937–1938	Only 27 built.	24	-
Unstyled AR	Row-Crop	1935–1949	New engine starting with tractors built 10/30/1940 and later. First unstyled AR built 8/9/1935. Last unstyled AR built 05/18/1949.	24	378
Unstyled AO (early)	Orchard	1936–1937	First production 05/22/1935. Replaced by AOS for four years. Last of first series built 10/05/1936.	24	-
AOS (streamlined)	Row-Crop	1936–1940	First streamlined AOS built 11/23/1936. Last styled AOS built 10/28/1940. Some had small curved flat fenders, others had fenders that cover top half of rear wheels.	24	-
Unstyled AO (late)	Orchard	1940–1949	First larger engine unstyled AO built 11/22/1940. Last unstyled AO built 05/17/1949.	-	-
AI	Industrial	1936–1941	First built 04/27/1936. Last built 06/18/1941. Only unstyled. Only 91 built.	25	-
Styled A (early)	Row-Crop	1939–1947	First early styled A built 09/12/1940. Last early styled A built 02/04/1947.	24	384 (1947)
Styled A (late)	Row-Crop	1947–1952	First late styled A built 03/31/1947. Last late styled A built 05/12/1952. Includes AN, AW, ANH, AWH.	24	-
Styled AN (early)	Row-Crop	1940–1947	First early styled AN built 09/11/1940. Last early styled AN built 02/04/1947.	24	335 (1939)
Styled AN (late)	Row-Crop	1947–1952	First late styled AN built 05/15/1947. Last late styled AN built 05/08/1952.	24	-
Styled AW (early)	Row-Crop	1940–1947	First styled AW built 09/10/1940. Last styled AW built 01/31/1947.	24	-
Styled AW (late)	Row-Crop	1947–1952	First late styled AW built 05/15/1947. Last late styled AW built 05/08/1952.	24	-
Styled ANH (early)	Row-Crop	1940–1947	First early styled ANH built 09/12/1940. Last early styled ANH built 01/31/1947.	24	-
Styled ANH (late)	Row-Crop	1950–1952	First late styled ANH built 07/10/1950. Last late styled ANH built 04/25/1952.	24	-
Styled AR (regular)	Row-Crop	1949–1953	First styled AR built 06/07/1949. Last styled AR built 05/06/1953. Kerosene engine 30 hp.	37 Gas	429
Styled AWH	Hi-Crop	1940–1947	First styled AWH built 09/11/1940. Last styled AWH built 01/15/1947.	24	-
Styled AO	Orchard	1949–1953	First Styled AO built 07/07/1949.	24	-

MODEL B

By Wayne G. Broehl Jr., from *John Deere's Company*

Deere's two new tractors were just the right sizes—the Model B was described as "two-thirds the size of the A," and company literature stressed that the A gave the pulling capacity of a six-horse team, the daily work output of eight to ten horses, with the B having the pulling capacity of a four-horse team and the daily output of six to eight horses. The B was made particularly as a one-plow tractor. It had all the features of the A—the adjustable wheel tread, the clearance, the excellent vision, etc. (and both were available with pneumatic rear tires as alternatives to the regular metal wheels). The B had a hydraulic power lift, and within a year of their introduction, both were also available with single front wheels (the AN and BN versions).

By 1936, there were eleven different versions of the company's three basic tractors—the A, B, and D. Both the A and B were also made with standard tread, there were orchard tractors available (with covered fenders for the rear wheels and no protruding stacks at the top of the tractor), and there were versions of the A and B that had adjustable front-wheel treads, as well as rear-wheel adjustments. As the success of the two new tractors sank in, company literature, such as the catalog of 1936, exultantly extolled the many new features and reaffirmed in aggressive advertising prose the efficacy of the two-cylinder tractor.

Probably no single stage in the entire history of the company's product development was any more important than this one. The appearance of the two new tractors in the depth of the Great Depression was a testimony both to the company's optimism about the future and the farmers' desires for and willingness to buy simple, trouble-free farm tractors. The abiding regard for and love of the two new models—the Model A and the Model B—continued all through the 1930s and '40s up to their final models in 1952 and left a residual of acclaim that made the models collector's items for tractor buffs down to today.

It is likely that the company would not have taken the steps to initiate these two new models had it not been for Charles Wiman. His longstanding love of both innovation and the engineering to back it up stemmed from his own academic training and his continuing enthusiasm. No armchair philosopher, he was always out in the field observing, and this was seen by everyone to enhance product innovation in the company at a time when it was desperately needed. The expenditures made in that uneasy period were substantial, particularly in light of the cuts in every other expense item in the company, both in factory and branch. It was a gamble by Wiman that was based upon faith and sound judgment, and its payoffs for the company in succeeding years were very great indeed.

MODEL B FIRST PRODUCTION

This is the first Model B produced, serial number 1000. The hood's leaping deer logo was used on only a very few early Model Bs, and the earliest Bs had the gas tank filler in the center of the tank underneath the steering shaft. *Keller Collection / Lee Klancher*

MODEL B FIRST PRODUCTION

Introduced in 1935, the Model B was a more compact, one-row sibling to the two-row Model A. Like the Model GP, the B would undergo many changes during the years it was produced. Very early Model Bs used a front pedestal that was attached using only four bolts.

SPARTAN ACCOMMODATIONS

Early farm tractor seats left a lot to be desired, and shifting the four-speed transmission was an awkward operation. More than 300,000 Model Bs were built between 1934 and 1952. This example is the first production model.

1935 MODEL B

The Model B was continuously improved throughout its production life. This early B has the optional adjustable rear fenders, and the four-bolt front bolster mount.

Vinopal Collection / Lee Klancher

1935 MODEL BN

The single-front-wheel Model BN is designed for work in narrow rows. Note that the early single-front-wheel Model Bs with four-bolt front pedestals were referred to as "Model B Garden Tractors." This Model BN has the later front bolster, with eight bolts fastening it to the frame. *Nelson Collection / Lee Klancher*

1936 MODEL BW

The Model BW has a wide and adjustable front end. This one was sold by the owner's grandfather at the John Deere dealership in Allen, Wisconsin. The owner was able to buy it back in 1999. *Holcomb Collection / Lee Klancher*

1936 MODEL BW-40

The BW-40 was a rare variant of the BW that could be narrowed to 40 inches wide. Only six were produced. *Keller Collection / Lee Klancher*

1938 MODEL BWH-40

The "40" designation came about because the first such model, the BW-40, could be narrowed to 40 inches at the rear wheels. That makes sense. This model, the BWH-40, could be narrowed to just over 42 inches, but the "40" designation was kept. The Model B model history is as convoluted as the family trees in *Game of Thrones*!

1940 MODEL BO

Produced from 1935 to 1947, the Model BO was equipped with large fenders, an under-slung exhaust, a slightly narrower width, and sculpted coverings over protrusions—all features designed for orchard work. The BO was a modified version of the Model BR, one key difference being differential braking, which allowed the user to make tight turns more easily, another beneficial feature when working an orchard. *Keller Collection / Lee Klancher*

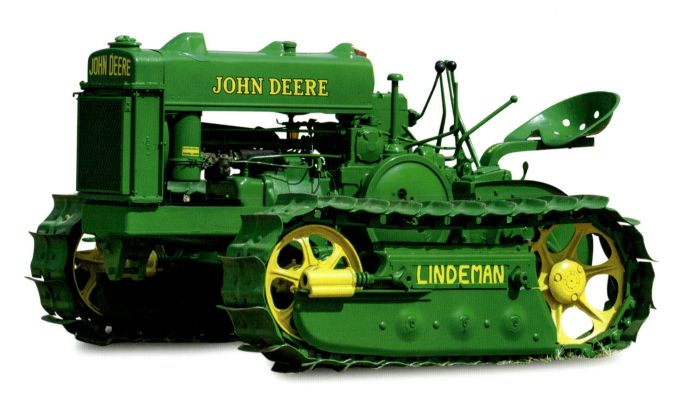

MODEL BO LINDEMAN

The Lindeman brothers of Yakima, Washington attracted Deere's attention by modifying a handful of Model GPs and a D to run on tracks. When the project went to production, the Model BO was used and built at the Lindeman facilities in Yakima. The production records were lost when the plant was closed, so precise production numbers are not known. An article in *Two-Cylinder* magazine used Deere factory records to estimate about 1675 Model BO Lindemanns were built. *Lee Klancher*

MODEL B DATA

Model	Type	Model Years	Notes	HP	Nebraska Test #
Unstyled B	Row-Crop	1935–1938	First production built 10/02/1934. Last production 06/14/1938.	16	305 (1938)
Unstyled BN	Row-Crop (narrow)	1935–1938	Replaced by Model M.	-	-
Unstyled BNH	Hi-Crop (narrow)	-	Only 65 built.	-	-
Unstyled BW-40	Row-Crop	-	Only 6 built	-	-
Unstyled BW	Row-Crop (wide)	1935–1938	247 built.	-	-
Unstyled BWH & BWH-40	Row-Crop	-	51 built of these two models.	-	-
Unstyled BR	Row-Crop (standard)	1936–1947	6,551 built.	-	232
Unstyled BI	Industrial	1936–1941	First production 04/01/1936. Last production built 02/27/1941. Only 183 built.	-	-
Early Styled B	Row-Crop	1939–1946	132,316 built.	21	366 (1941)
Late Styled B	Row-Crop	1947–1952	89,334 built.	28	380 (gas), 381 (all fuels)
Styled BWH	Row-Crop	1939–1946	268 built.	-	-
Styled BWH-40	Row-Crop	1939–1946	Replaced by Model M. Only 12 built.	-	-
BO	Orchard	1936–1947	5,035 built.	-	-
BO Lindeman	Orchard	1936–1947		-	-

Theo Brown spent the first weeks of 1933 sketching a lift for a new tractor. He wrote that the HX was designed to be 65 percent the size of the forerunner to the Model A, and that they hoped to have prototypes running by mid-April. By early February, Brown had come up with a new swinging drawbar design and a ratcheting engagement for the power lift.

The following October, Brown was testing the Model HX in the field with his implements. He was pleased with the results and pasted a number of photographs of the tractors at work into his journal. On October 30, he added a chart showing the number of patent applications filed by each of the major tractor makers between 1931 and 1933. John Deere had filed more than 234 requests—an amazing 41.5 percent of all requests filed.

Brown's power lift was a sensation. On December 1, 1933, he wrote, "IH seems to be trying to keep away from the power lift on the Farmall as they must realize that we have the idea so well patented. I'm hoping we will capitalize the power lift and get real business from the monopoly we ought to have with it."

A few weeks later, as drizzle fell outside, the experimental group presented to the chairman. The presentation included Elmer McCormick's conclusions regarding the GX and the HX, now known as the new Model A and B, respectively. After a few days of Christmas parties and "blissful relaxation" with his wife, Brown wrote a few lines about his approach to design: "In thinking ahead as to what the tendency in implement design will follow, it seems as tho [sic] it might be wise to go around somewhat and talk

PATENT PAYMENT

Theo Brown's lift patent eventually led to IH paying John Deere for every single lift they sold that used a similar design.
Theo Brown Diaries / WPI

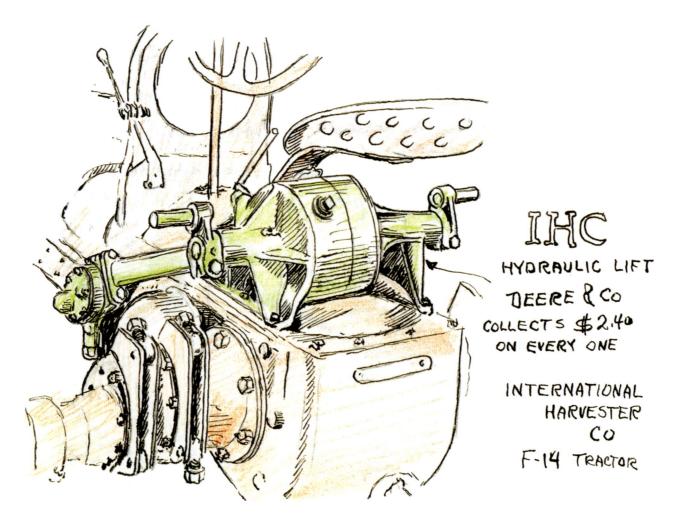

IHC
HYDRAULIC LIFT
DEERE & CO
COLLECTS $2.40
ON EVERY ONE

INTERNATIONAL
HARVESTER
CO
F-14 TRACTOR

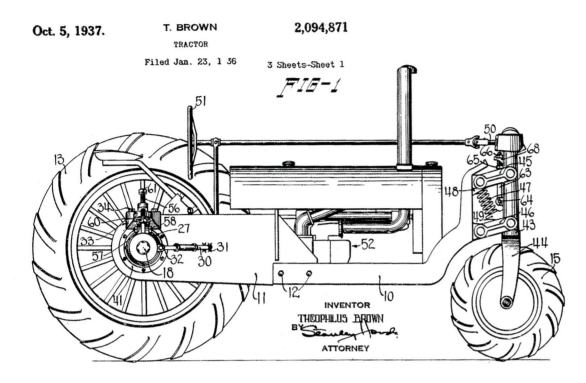

Oct. 5, 1937.   T. BROWN   2,094,871

TRACTOR

Filed Jan. 23, 1 36   3 Sheets-Sheet 1

FIG-1

INVENTOR

THEOPHILUS BROWN

BY Stanley Houck

ATTORNEY

THE AWFULEST BROWN

Theo Brown was awarded 158 patents during his long career at Deere. This patent is for a lockable suspension system on a tractor, which would allow a smooth ride on the road and, when locked out, good performance in the field. Brown's prolific production of patents prompted IH to dub him "The Awfulest Brown." *US Patent*

PATENT CHART

Theo Brown kept detailed records on all kinds of things, including this incredible number of patents created by John Deere in the 1930s. *Theo Brown Diaries / WPI*

to farmers and find out what they think they would like to do with machinery etc. that they can't do now. In visiting I might be able to get some new hunches and develop to a point where we could get patent applications in. Now we are using ideas on which I applied for patents five years ago. It is much better to have others try to copy us than to have to try to copy the other fellow."

This philosophy would mark Brown's work. His creativity was fueled by conversations with men of the day who worked long hours with tractors and implements, and his sketches and thoughts resulted in bits and pieces on most John Deere tractors created between 1918 and 1952.

Also note that Theo's work on the power lift generated significant income for John Deere. On May 25, 1937, he wrote in his diary, "Silver told me they [International Harvester] are making 300 mechanical power lifts a week and their schedule calls for 18,000 for the year. This at the royalty of $4.80 each which they pay Deere & Co. would mean $86,400.00 for the year.

The next time you examine a power lift on a John Deere of this vintage, know that the words of hardworking farmers were catalysts for the steel in your hands.

NUMBER OF FARM EQUIPMENT PATENTS ISSUED
INDICATING RELATIVE ACTIVITIES OF COMPANIES BY PERCENTAGES IN RED FIGURES.

	1930	%	1931	%	1932	%	1933	%	1934	%	1935	%	1936	%	TO DATE 1937	%	TOTAL	%
DEERE	33	29	60	35	110	46	104	50	89	50	40	34	24	23	12	22	472	40
IHC	52	45	69	40	82	35	54	26	57	32	53	45	51	49	28	52	446	38
CASE	10	9	19	11	26	11	23	11	15	8	13	11	8	8	3	6	117	10
OLIVER	8	7	13	8	10	4	8	4	6	4	5	4	6	6	3	6	59	5
ALLIS-CHALMERS	7	6	4	2	1	0	11	5	6	4	3	3	4	4	3	6	39	3
MASSEY-HARRIS	2	2	3	2	6	3	5	2	4	2	1	1	6	6	2	4	29	2
M-M	3	2	3	2	3	1	4	2	1	0	2	2	4	4	2	4	22	2
TOTALS	115	100	171	100	238	100	209	100	178	100	117	100	103	100	53	100	1184	100

NOTE   42 PATENTS ISSUED TO IHC ON MILK EQUIPMENT
CRAWLER TRACTORS AND TRUCKS NOT INCLUDED IN ABOVE LIST

MODEL BI INDUSTRIAL

Deere engineers such as Theo Brown spent part of their time out in the fields asking farmers how to improve their machines. Opportunists like his colleague McCormick were finding new ways to sell their tractors. One of the potential venues identified was heavy industry.

Tractors with low profiles, high gearing, and attachment points were useful in factories for everything from operating pallet lifts to towing railcars and airplanes and cutting the grass. Such machines had been at work at Deere's own factories for many years. Marketing them to others was only logical.

One of these industrial models was a conversion of the Model B known as the Model BI. The major engineering change was a higher gear ratio in the rear end, which allowed the machine to scoot around the factory or town at a reasonable pace.

Deere & Company built two experimental versions of the Model BI in February 1936, and the first production Model BI arrived a few months later. The machines were typically painted yellow with a few sold in other colors. The Model BI was also sold through Caterpillar dealerships and other industrial equipment outlets. While these partnerships had promise, they didn't deliver. The sales goal for the Model BI was 500 units per year—actual production for 1936 was 183 machines.

1937 MODEL BI

The Model BI's front axle is set farther back than on a Model B Standard, shortening the wheelbase. The front end also features pads and tapped holes for mounting equipment, while the low-exiting exhaust is distinct to the industrial model, giving the tractor better clearance for use inside factories and warehouses.
Keller Collection / Lee Klancher

1937 MODEL BI

While most Model BIs were painted yellow, a few were ordered in specialty colors. This 1937 tractor was ordered by the city of Waterloo, Iowa, where it pulled the

fire department's ladder truck and was used to mow grass and pump water. The Model BI was produced from 1936 to 1941. The last one was built in February 1941.

1937 MODEL BI

The BI's seat is a special unit that is adjustable fore and aft. The Model BI features a beefed-up rear axle, with heavier bearings and shafts than those used on the Model B Standard.

In the first part of the twentieth century, Industrial Design transformed the way products were created. Designers such as Teague, Raymond Loewy, and Henry Dreyfuss led high-profile firms that created beautiful designs for everyday objects.

The best of the designs combined aesthetics with utility, making objects such as telephones, thermostats, and refrigerators pleasing to the eye and also easy to use. These "styled" products sold much better than their clunky counterparts, and this group of designers became highly sought after by manufacturers of all kinds.

One of the most famous was Henry Dreyfuss, who was perhaps best known for his work on the streamlined Hudson J-3a steam locomotives that pulled the New York Central Railroad's *20th Century Limited*.

By the 1930s, a number of tractor manufacturers were experimenting with streamlining, bright paints, and other styling exercises. While engineers tried their hand internally—see previous piece on the Deere Model AOS—many of the large tractor manufacturers hired some of the era's top industrial designers to help with their line. International Harvester commissioned Raymond Loewy, and Brooks Stevens penned models for Allis-Chalmers and Minneapolis-Moline.

Deere raised the bar beyond the previous efforts of in-house engineers when it hired Henry Dreyfuss. His firm turned out to be a perfect fit for the company.

One of Dreyfuss's favorite stories was from his early career, when a movie studio sent him to Sioux City, Iowa, to investigate why their brand-new movie palace

there was drawing tiny crowds. Dreyfuss lowered ticket prices, ran triple features, and gave away dish sets, but still found that locals preferred an older theater down the street. He decided to spend three days standing quietly outside the new theater, watching people walk past. Afterward, he had the lush, red carpet in the lobby replaced with a plain rubber mat. Ticket sales exploded. The problem, Dreyfuss discovered, was that local farmers and townspeople hadn't come through the door for fear of dirtying the carpet with their boots.

HENRY DREYFUSS (1904-1972)

One of the pioneers of industrial design, Henry Dreyfuss was known for designing iconic American products, including the Princess and Trimline telephones, Hoover vacuum cleaners, and the 20th Century Limited locomotive. He and his firm designed John Deere tractors from 1937 to well into the twenty-first century.
Henry Dreyfuss Archive, Cooper Hewitt, Smithsonian Design Museum

The son of immigrant parents, Dreyfuss was born in New York City and from 1920 to 1922 attended the Ethical Culture Society's private high school. He went on to study stage design under Norman Bel Geddes. Geddes was one of America's first industrial designers, and he taught Dreyfuss how to methodically transform visualizations into physical forms. Working as set designers also taught both men how to create designs that met clients' needs and tastes, a key aspect of their success.

By 1925, the twenty-one-year-old Dreyfuss had achieved modest success as a stage designer in New York and determined that his future was as an industrial designer. As the 1930s progressed, Dreyfuss designed corporate airplane interiors for William K. Vanderbilt and others. These dramatic stylings led to coverage in *Architectural Digest* and the assignment to redesign the locomotive and cars for New York Central's *20th Century Limited*, which traveled between New York and Chicago. When the streamlined train debuted in 1936, it was heralded as a breakthrough in railroad design.

Some of the Deere engineers at Waterloo Tractor Works eventually convinced Charles Stone, the head of manufacturing, to bring in industrial designers to restyle their tractors. In 1937, Stone sent chief engineer Elmer McCormick to New York to speak with Dreyfuss about this idea.

Dreyfuss, who believed that replacing artifice with function was one of his crowning achievements, proved a good fit for Deere. He accepted McCormick's proposition and went to work with Deere almost immediately. The look of John Deere tractors would never be the same.

The first assignment for Henry Dreyfuss Associates (HDA), his design firm, was redesigning the Model A and B tractors. The changes suggested by HDA not only dramatically improved the appearance of the machine, but also enhanced visibility, safety, and the cost-effectiveness of production. Deere had HDA redesign nearly its entire line of tractors, and the "styled" farm tractor was introduced to the market for the 1938 model year.

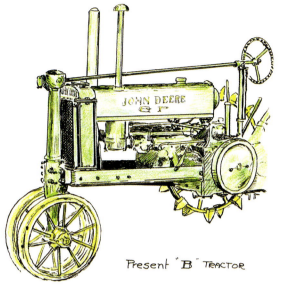

Present "B" Tractor

New Styling "B"

An Idea for Tractor Styling

THEO'S DIARY

John Deere engineer Theo Brown was captivated by the new design of the Model A and B, and sketched the dramatic change in his diary on March 3, 1938. *Theo Brown Archives / WPI*

The tractors sold well and their appearance garnered accolades for decades. Other tractor makers would also turn to well-known industrial designers. The men who styled the other colors were great designers, but none would have meshed with Deere as well as the tidy and thoughtful Henry Dreyfuss, who was known for his elegant style, impeccable ethics, and ever-present brown suit.

THEO'S IDEA

Several weeks after sketching the Dreyfuss design, Brown sketched his own idea for a styled Deere tractor. Page dated March 18, 1938. *Theo Brown Archives / WPI*

The philosophy of functional, simple, and stylish appearance that characterized HDA was a wonderful match for the methodical, research-driven approach of Deere. The partnership forged in New York in 1937 continued well into the twenty-first century.

STYLED MODEL A & B

By the mid-1930s, the industrial world was enamored with design. Styled products such as George Grant Blaisdell's timeless Zippo lighter, Walter Dorwin Teague's Brownie camera, and Henry Dreyfuss's Bell telephone were everywhere.

Men like Harley Earl sketched new designs that would transform automobiles from self-propelled buggies to rolling sculptures. The mechanical exteriors of locomotives were sheathed in sheet metal. Raymond Loewy's buses prowled the streets. The streamlined results became timeless pieces of American culture.

For 1939, Deere's Model A and B were rolled out with the styling of Henry Dreyfuss applied. That machine was produced through February 1947,

STYLED MODEL AR

The Model AR and the orchard version, the AO, were built as styled machines in June 1949. Production continued until May 1953, when the model was replaced by the Model 60. This one was photographed in front of the Plaza Theater in Lamar, Missouri. The Art Deco theater is an appropriate backdrop for the lovely styling on the AR.
Purinton Collection / Lee Klancher

EARLY STYLED MODEL B
You can identify late versus early Model As and Bs by the frame. Early styled have pressed-steel frame (like this example) and later ones are welded. *Lee Klancher*

STYLE COMES TO DEERE MODEL B

The vision of Henry Dreyfuss first came to life at Deere in the styled 1939 Model A and B. This farmer is hauling oats with one on August 6, 1948. The seven slots in the front grille are an indicator it is a Model B—the Model A and G have eight slots in the grille. *National Archives*

when Deere needed to apply another change to the Model A.

From 1947 to the end of production in 1952, Deere Model As and Bs had a series of significant updates made.

The one that *Two-Cylinder* magazine credited as the most significant was the newly added gasoline engine option. Previous versions were designed to burn distillate or other low-cost fuel, and Deere had heavily marketed these feature for decades.

In addition to the new "Cyclone" gasoline engine—which offered more horsepower than the distillate option—the updated Model A and B included electric starting, rubber tires, and lights as standard equipment, a new seat, and a host of other improvements.

As the more staid models trickled out the door, Deere executives snapped photographs in the field with Teague's sleek Brownie camera and drove to work in their swoop-fendered Buicks and Chevrolets. The sway that style could hold over the floor managers and corporate buyers who purchased street sweepers and tugs was most likely minimal, but the impact of industrial design on society at large was clear every time someone picked up the phone or stepped out the door.

1952 MODEL A HI-CROP

1952 Model A Hi-Crop.
Mecum Auctions

See page 63 for styled Model A tractor data, and page 75 for styled Model B tractor data.

LATE STYLED MODEL B ALL-CROP HIGH-CLEARANCE UNIT

Manufactured by H. Stewart of Salinas, California, this late styled Model B was modified to offer super high clearance. *Keller Collection / Lee Klancher*

1950 LATE STYLED MODEL B

Mecum Auctions

One of the little-known sidebars in early Deere tractor development is the Sklovsky Three-Wheel Tractor. Deere was very interested in developing a one-plow tractor for small farmers, and in fact was quite progressive on this front. While a few manufacturers were experimenting with such small machines, John Deere management was ahead of the curve to be looking in this direction in the early 1930s.

Max Sklovsky was one of the key John Deere engineers, and Deere built and tested one of his tractor designs in 1915. That machine never made it past the experimental stage.

In the hot July of 1932, direction from management prompted the engineering team to build and test a Sklovsky design tractor. Engineer Theo Brown noted in his diary that the design specified a 2,100-pound machine with a single rear drive wheel and two front wheels spread 68 inches apart. Brown was skeptical from the beginning.

On July 18, 1933, Brown wrote, "I told Charlie Wiman, who was very enthusiastic about this idea, that I did not think an outfit like this would steer satisfactorily for a cultivator in turning at ends of hill. Made a rough model full size which we had in the field. Charlie Wiman, Frank Silloway, Max [Sklovsky], and Lesser [Nathan Lesser, Sklovsky's nephew] present. The demonstration not particularly encouraging. Charlie said he thought I had put my finger on the sore spot."

FIELD TESTING

These images show the Sklovsky machine in the field in March 1933. Theo Brown wrote that Charlie Wiman, Frank Silloway, Max Sklovsky, and Nathan Lesser were present.
Theo Brown Diaries / WPI

Brown wrote more about the tractor in March 1933, when the Sklovsky tractor was tested in the fields. His first entry on March 16 shows mild encouragement. "Spent all day working in field with working model of Max Sklovsky's tractor," Brown wrote. "Worked better than any of us expected it to." The Sklovsky prototype was tested on March 17 and 18, and again March 28 and 29.

Brown wrote a report summarizing the Power Farming Committee's findings about the tractor, comparing it to the Deere experimental HX tractor (which would later become the Model B). Brown gives kudos to Sklovsky, Lesser, and the Harvester Works staff for building a good quality full-size and operational machine, calling it a "most creditable piece of work."

Brown's report includes skepticism of the viability of the configuration. "A number of tractors have been built with a single drive wheel but none have been successful. Possibly the tractor that most closely approximates the Sklovsky design is that of the IH motor cultivator built in 1917," Brown wrote. "Walter Silver followed several of these tractors at the time and his report is to the effect that these tractors were a failure because of a lack of stability and traction."

"The Committee feels that the only advantage it can find in the Sklovsky design does not meet the requirements of a one-plow general purpose tractor."

On March 29, Elmer McCormick, Paradise, Max Sklovsky, White, and Theo Brown met to discuss the tractor, and decided to end development of the Sklovsky tractor and focus their efforts on the HX tractor.

Brown put the finishing touches on the report on the thirtieth, and wrote, "Finished report on Sklovsky tractor and after some hesitation Max signed it with the rest of us."

Sklovsky continued his work developing a small machine. He would work on another failed, smaller prototype tractor in 1938. Before that, his son, Ira Maxon, would develop the machine that would finally satisfy Deere's hunger for a one-plow machine.

EARLY DESIGN

This design was created by John Deere engineer Max Sklovsky, and was ultimately scrapped. The work undoubtedly factored into the Model, which Sklovsky collaborated on with his son, Deere engineer Ira Maxon. *Theo Brown Diaries / WPI*

MODEL L

The 1925 U.S. Census showed America's roughly 6 million farms used just over 505,000 tractors. That left 5.5 million farms without tractors.

According to U.S. Census data, 60 percent of American farmers worked under 100 acres and 38 percent had less than 50 acres. That meant roughly 3.3 million small farms were a ripe market for a small tractor that could replace a horse or mule.

Deere had been chasing this market for quite some time, and the company created the Model L for this large group of small farmers. Deere CEO Charles Wiman had been fascinated with the odd Sklovsky Three-Wheel Tractor—see sidebar—and the company continued to look for a viable tiny machine.

With Theo Brown and Elmer McCormick nearly overwhelmed with their design responsibilities, the task of creating this small tractor was shifted to Ira Maxon of the Moline Tractor Division. He brought in former Deere engineer Willard Nordenson to help his thinly funded division create the machine.

Nordenson designed a machine that utilized a short tube-steel frame, a foot-operated clutch, and an outsourced engine. The machine, created with a tiny budget, provided solid performance and developed from the experimental Model Y to the early production Model 62 to the Model L.

**BROWN SKETCH,
FEBRUARY 23, 1939.**

Theo Brown Diaries / WPI

No 7 Mower L Tractor

MODEL 62

This small model was originally produced in 1936 as the Model Y prototype and later as the Model 62.

John Deere Archives

MODEL 62, SERIAL NUMBER 78

This Model tractor was developed by Ira Maxon, the son of Max Sklovsky, with help from Willard Nordenson. Very early Model Y prototypes used a Novo engine; production Model 62s used a Hercules. *Keller Collection / Lee Klancher*

MODEL L, SERIAL NUMBER 79

In 1938, the Model 62 designation changed to the Model L. The unstyled Model L was replaced by the styled version for 1939. *Keller Collection / Lee Klancher*

STYLED MODEL L

Model Ls built after August 1938 received the Dreyfuss styling treatment. The first styled Model L is serial number 625000. The early Model L was powered by a Hercules NXB vertical twin-cylinder gas engine. Later models used a 66-ci John Deere vertical twin. *John Deere Archives*

MODEL LA

One of the most interesting, styled John Deere models is the LA, which had significant input from Henry Dreyfuss and company. HDA became involved with the project in November 1937. They drew the distinctive lines of the styled Model LA and offered feedback on dozens of details.

In the well-researched account of the Model L's evolution in his book *John Deere: A History of the Tractor*, author Randy Leffingwell wrote of the LA, "As each new element came under scrutiny, an assortment of paper went by airmail to and from designers and engineers. Seat shape and gear-change lever configurations and the shape of the perforations in hood and grille sheet metal provoked suggestions and alternatives. Dreyfuss's staff and Nordenson's engineers drew and redrew."

One key member of the Dreyfuss staff was Roland Stickney, a highly respected designer best known for his work on the building design for Rockefeller Center and on Duesenberg, Chrysler, and Lincoln automobiles. Stickney transformed the Dreyfuss concept sketches into finished drawings. Many of the sketches done for Deere & Company are credited to Stickney.

The machine created by the engineering efforts of Maxon and the design input of HDA proved timeless. More importantly, the close interaction between the design firm and the engineers would become an integral part of the design process at Deere & Company.

MODEL LA

The Model LA was a variant of the Model L with a more powerful engine, cast wheels, and—on most examples—round bar rear frame members. This one is skidding logs in Maine. *National Archives*

1941 MODEL LI FIRST PRODUCTION

The Model LI was another "L" variant produced from 1941 through 1946. This Model LI is the first production LI. It was shipped to John Deere Harvester Works in Moline and used for various jobs at the plant. *Keller Collection / Lee Klancher*

Model	Type	Model Years	Notes	HP	Nebraska Test #
MODEL L & LA DATA					
62	Row-Crop	1937	Replaced experimental Model Y.	-	-
Unstyled L	Row-Crop	1937–1938	Replaced Model 62. Hercules engine.	10	313
Styled L	Row-Crop	1939–1946	Both Hercules and Deere engines used. Approx. 400 had Hercules engine.	-	-
Unstyled LI	Industrial	1938–1940	SNs interspersed with Model L. Less than 70 LIs with Hercules engine.	-	-
Styled LI	Industrial	1941–1946	Unique SNs for this model starting in 1941. Deere engine.	-	-
LA	Row-Crop	1941–1946		14	373

MODEL G

As America climbed out of the great depression, Deere expanded its line. One addition was a larger sibling to the A and B. Developed as the KX in 1936, the resultant Model G was authorized for production in January 1937.

The machine emerged as an experimental model in May of that year, with fifty units built in the ensuing few months. The tractors were sent out for testing, a few going to Kentucky and Missouri, with most of the rest shipped to Minnesota, Iowa, and Illinois.

In these tests, the Model G engines would run hot enough to burn valves, particularly in southern climes. In response, the engineering department developed and tested a taller radiator. In addition, they redesigned the radiator shutter, fan, and shroud, and even the hood and fuel tank. But the solution proved to require more than additional cooling: The exhaust valves were seated deeper in the head, allowing them to cool better and thus reducing their operating temperature.

Once Deere had the redesign in place, it changed the production models and recalled the roughly three thousand "low-radiator" models that had been built and sold. Most of these early machines were brought back in for the retrofit, but a few recalcitrant farmers, perhaps working in northern climates that posed no overheating issues, neglected to bring in their machines. Collectors today prize the few remaining low-radiator Model Gs.

During World War II, the Model G specifications changed, and it was designated the GM. After the war ended, the G returned. In 1950, the Model GW and GH were added to the line.

MODEL G

The Model G experimentals were dubbed the KX, and records show the four were shipped in 1936 and early 1937. The model designation would have logically been the Model F, but Vice President of Sales Frank Silloway wanted to avoid any confusion with IH's line of F-series Farmalls and this tractor became the Model G. About 64,000 production units were built from 1938 to 1953. The Model G produces 27.6 drawbar horsepower and is equipped with a power lift and 532 rpm power take-off.

Keller Collection / Lee Klancher

MODEL G "HIGH RADIATOR" EXPERIMENTAL

The unstyled Model Gs were first built in 1937, and the early models had a four-speed transmission and a smaller radiator. This tractor is serial number G1045, the first unstyled Model G fitted with an experimental high radiator. The last unstyled Model G was built on December 22, 1941.

A design sketch for the Model GM. The eight slots in the front hood identify this as either an A or G, and the exhaust and intake being orientated side-by-side in the hood indicates this is a concept sketch for a Model GM. The first of the new improved Model GMs was built in February 1942. When the machine was styled, it was treated to an updated six-speed transmission. After the war, the GM was replaced by the late-styled Model G. Production of the late-styled Model G began in March 1947. The easiest way to identify a late-styled G is the cushioned seat; the GM had a steel seat.

Henry Dreyfuss Archive, Cooper Hewitt, Smithsonian Design Museum

MODEL G DATA					
Model	**Type**	**Model Years**	**Notes**	**HP**	**Nebraska Test #**
Unstyled G	Row-Crop	1938–1941	First production 05/17/1937 (SN 1002). Last production 12/22/1941.	36	295
GM (styled)	Row-Crop	1942–1947	Replaced G during World War II. First production 02/21/1942; last 03/10/1947. Steel uncushioned seat.	-	-
Late Styled G	Row-Crop	1948–1953	34,474 built. Cushioned seat distinguishes it from Model GM.	38	383
Styled GN	Row-Crop (narrow)	-	1,571 built.	-	-
Styled GH	Hi-Crop	1950–1953	237 built.	-	-
Styled GW	Row-Crop	1947–1953	4,785 built. Last production 02/1953.	-	-

MODEL H

Designed to be less expensive and smaller than the Model G, the Model H began life in 1938 as the experimental OX, which was unstyled for a short period and, when the Model A and B production models were released, the experimental OX was also styled.

A run of 104 Model Hs was built in the fall of 1938 and, according to *Two-Cylinder* magazine, most of these were scrapped and only a handful of those survived. Regular production of the model began in January 1939.

The Model H had less than ten drawbar horsepower when tested at Nebraska, making it capable of pulling a single 16-inch plow or two-bottom 12-inch plow.

Production was paused during World War II, when material shortages made production difficult.

MODEL H LIFT

Theo Brown sketch dated February 3, 1941. *WPI*

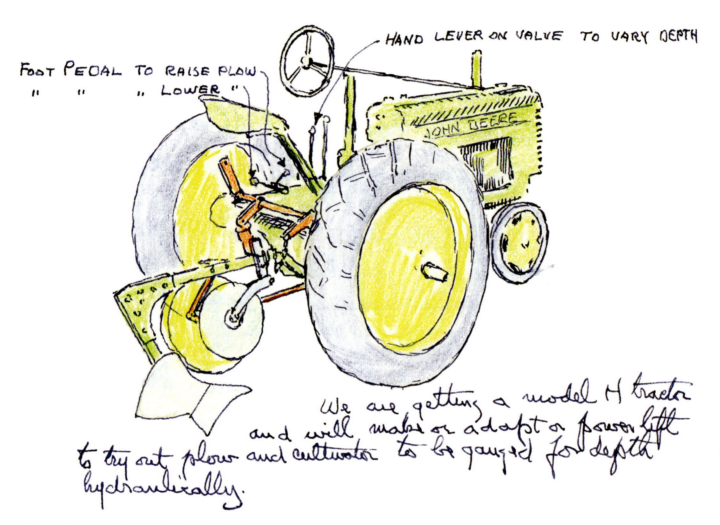

HAND LEVER ON VALVE TO VARY DEPTH

FOOT PEDAL TO RAISE PLOW
"   "   " LOWER "

We are getting a model H tractor and will make or adapt a power lift to try out plow and cultivator to be gauged for depth hydraulically.

MODEL H DATA					
Model	Type	Model Years	Notes	HP	Nebraska Test #
Styled H	Row-Crop	1939–1947	First built 10/29/1938 (scrapped). Last built 02/06/1947. Kerosene engine.	15	312
Styled HN	Row-Crop	1939–1947	First built 11/10/1939 (experimental). Kerosene engine.	-	-
Styled HWH	Row-Crop	1939–1947	First production 03/06/1941; last 01/29/1942. Kerosene engine.	-	-
Styled HNH	Row-Crop	1941–1942	Kerosene engine.	-	-

MODEL HNH

The first Model H preproduction model was built on October 28, 1938, and shipped to the University of Nebraska for testing. Regular production of the Model H began on January 18, 1939. The early experimental versions of the Model H were known as the Model OX. A few of these experimental models were unstyled, but all production models were styled. *Keller Collection / Lee Klancher*

MODEL HNH

The Model HNH is a narrow version of the Model H, which began development in the fall of 1938. This is serial number 41760, which was built on December 11, 1941, and shipped to Los Angeles. Only thirty-seven Model HNH tractors were produced between March 1941 and January 1942.

MODEL HNH COCKPIT

The Model HNH is equipped with a special rear axle housing, which provides additional ground clearance. The machine was developed to meet the needs of California farmers. Dreyfuss and the design team at HDA redesigned the dash of styled John Deeres, making them more handsome and easier to read.

TRACTORS AT WAR

War in the twentieth century increased the demand for farm equipment due to labor shortages on farms and the need for unmolested land to feed the world. This poster was put out by the U.S. Office for Emergency Management in the latter part of WWII. *National Archives*

By mid-1940, Deere's pace had quickened; by December of that year, it was making war material in the Welland plant, to be delivered to the Canadian government. By March 1941, the company was the prime contractor for the fabrication and production of transmissions and final-drive units for the M3 medium tank. The work was done in the Waterloo plant, but under the aegis of a separate company called the Iowa Transmission Company.

After this first large defense contract, the company joined the rest of American industry over the following months of 1941 to gear up for greater defense produc-

tion. In March 1942 a major new assignment came; Deere was to be a subcontractor to the Cleveland Tractor Company in the manufacture of MG-1 military tractors for the United States armed forces. Heavy tractors (weighing approximately seven tons) with track-type traction, they were intended primarily for use at airports in servicing and pulling aircraft. The Cleveland Tractor Company was one of the important manufacturers of track-laying tractors, under the trade name Cletrac. For some time, Wiman had seemed concerned about how Deere would keep its organization together in the face of war demands (in several previous board meetings he

FOOD COMES FIRST

had articulated his own personal worries about this), and now he apparently saw a solution in the Cleveland Tractor link. His idea (as espoused in the board meeting of April 28, 1942) was a merger of the two companies! The notion was debated vigorously in the meeting amid much opposition.

Almost immediately, however, the Cleveland Tractor merger proposal was sidetracked. Charles Wiman's career had built him a wide reputation among government administrators; now he was asked to come to Washington as a colonel in the Ordnance Corps of the United States Army, to work directly in the tank and combat vehicle division. Wiman felt that he had no choice. He resigned as president of Deere & Company and left immediately for Washington. Burton Peek was elected president of the company, to hold responsibility while Wiman was gone. This sudden change in top management doomed the Cleveland Tractor merger proposal, since Wiman had been its chief advocate.

Deere also engaged in a wide range of other defense efforts—contracts for ammunition (75-millimeter and 3-inch shells), subcontracts to provide parts to aircraft manufacturers, and the building of mobile laundry units. Deere assembled the laundry units, which were pulled as semitrailers by large trucks, to accompany combat troops and thus reduce the age-old problem of war, the vermin that cause so many diseases. The Wagon Works of the company also made various cargo units and just about all of the factories of the company were devoted in a major way to some war production. By the end of the war, more than 4,500 employees had entered military service; even a "John Deere" ordnance battalion had been formed, recruited mainly from employees of factory, branch-house, and dealer organizations, and had served in the European theater of operations.

As shown by the figures for 1944, the market shares of the seven long-line companies declined during the war years, due to the limitation orders. By 1948 all seven had made strong recoveries; the four smaller of the seven registered proportionally larger gains from 1944 than did the three large companies that initially were parties to the suit. The two smallest, Massey-Harris and Minneapolis-Moline, had not only more than doubled their 1944 market shares, but they had also exceeded their positions in 1936 by substantial margins. International Harvester was the least successful in making a comeback, though it continued to lead the industry.

WAR EVENS THE PLAYING FIELD

Market Shares of the Seven Full-Line Agricultural Machinery Manufacturers, 1944–1948 (percent of total industry sales)

	1944	1945	1946	1947	1948
J. I. Case	3.7	4.4	3.8	5.3	7.0
International Harvester	15.8	17.6	22.3	23.4	22.8
Deere	9.6	9.9	15.1	14.1	15.3
Allis-Chalmers	3.2	4.3	3.5	6.3	6.8
Oliver	2.1	2.9	4.3	3.9	4.2
Minneapolis-Moline	1.7	1.9	3.1	3.3	3.6
Massey-Harris	1.1	1.3	2.9	2.8	3.8
Full-line companies' share of industry sales	37.2	42.3	55.0	59.1	63.5

Source: United States v. J. I. Case Co., 101 Fed. Sup. 856.

MODEL M

MODEL M

The new-for-1947 M was an all-new one-row tractor with 20 horsepower, hydraulics, PTO, electric starting, and clean Dreyfuss styling. The first one was delivered to the ranch of company leader Charles Deere Wiman on April 1, 1947. *Vinopal Collection / Lee Klancher*

The diminutive Model M played a number of cards with its appearance in the 1947 Deere lineup. Created by the same engineering team that brought you the Model L, it replaced that little tractor as the small utility rig in the line. The Model H was also pulled from the line to make room for the M, and the M had more power and capability than the H.

The M also was widely considered a Ford 9N fighter, and was the first tractor built at Deere's newly opened Dubuque manufacturing plant.

With variations for industrial and orchard work, cultivation, and even with tracks, more than eighty-eight thousand Model Ms were built from 1949 to 1952.

MODEL MC CRAWLER

The first MC crawler rolled off the line at Yakima Works—the former Lindeman plant owned by Deere—on March 3, 1949. *John Deere Archives*

Model	Type	Model Years	Notes	HP	Nebraska Test #
MODEL M DATA					
M	Row-Crop	1947–1952	Built at Dubuque Tractor Works. First production 04/01/1947. All-fuel or gas engine.	29	387
MC	Crawler	1949–1952	First production late 1948; last 09/1952. 10,509 built.	29	448
MT	Row-Crop (tricycle)	1949–1952	First production 12/1948; last 09/1953. 30,472 built.	29	423
MI	Industrial	1949–1952	First production 11/1949; last 08/1952. 1,032 built.	-	-

MODEL R

The Model R had one of the longest development cycles of any John Deere tractor. When it was released to the public in 1949, the tractor had spent fourteen long years in development and more than sixty-six thousand hours being field-tested.

Large cylinders aren't terribly conducive to diesel ignition. Deere's job would have been much simpler had it been willing to build a four-cylinder diesel engine. The company was not. The resulting Model R was the most powerful farm tractor built by Deere up to that time and possibly the most reliable, not to mention one of the best-looking.

According to *Two-Cylinder* magazine, Deere engineering began their work on the Model R diesel engine in 1935, when several engine designs were tested and scuttled. It took them until June 1940 to have an engine that worked well enough for Deere

management to authorize building and testing the rest of the machine.

An experimental version of the Model R—the Model MX—was first tested in 1941. Another eight years would be spent developing the machine for production. Introduced in 1949 and produced until 1954, the production Model R featured a diesel engine praised for its power, fuel economy, and reliability.

The Model R's design was heavily influenced by Henry Dreyfuss's "form follows function" philosophy. When testing showed that the large cooling fan required to keep the big engine running at the proper temperature gathered debris on the grille, Dreyfuss designed the angle of the grille corrugations so that farmers could easily sweep off the chaff with their hands.

MODEL R

The Model R seen here was sold to Paul Ashauer on June 2, 1953, at the Keller John Deere Dealership in Wisconsin. The tractor was too large to truck, so an employee drove it to Ashauer's farm. Ashauer meticulously cared for the R, and most of the original parts are intact. The Model R's production years were 1949 to 1954, and it was tested at Nebraska in test number 406. *Keller Collection / Lee Klancher*

CHAPTER THREE

THE ART OF REFINEMENT

NUMBERED TWO-CYLINDER MACHINES

"These new tractors were designed, in a sense, by our farmer customers," —Deere general sales manager Lyle Cherry, when he introduced the new Model 50 and 60 on June 11, 1952, *Two-Cylinder* magazine

CUSTOM-BUILT MODEL 60

The Model 60 was not sold by John Deere with four-wheel-drive, but this handmade custom shows how one may have looked. *Sweeney Collection / Lee Klancher*

The mid-1950s were turbulent times at Deere & Company. Longtime company president Charles Wiman passed away in 1955, opening the door for change.

Wiman's dedication to research and development and product excellence, along with his fiscal conservatism, had vaulted Deere into the industry's second-place slot behind tractor giant International Harvester. This was an incredible leap. Deere had to start from scratch with tractors and compete with a behemoth agricultural company formed by combining under one "company" more than 70 percent of *all agricultural manufacturers*—and possibly as much as 85 percent. The International Harvester Company was a dominating trust—such mergers are now illegal and IH would be sued and required to divest of some properties in 1917. To make the market more difficult, Henry Ford built tractors and did so with deeper pockets than any of the rest and little regard for profit.

For the independent Deere to grow its presence in tractors from literally nothing to being a strong number two—and lead in the number of patents created—is one of the most remarkable examples of smart business management and engineering prowess in the twentieth century.

So while we must give Wiman his due, at the same time we have to acknowledge that throughout the 1940s, his management style indicated he was content with the John Deere market position. They were the ethical leaders, cleverly using old technology—the two-cylinder engine—to carefully craft outstanding farm machines. This is a respectable position, but it was not that of the industry leader.

While Wiman carefully managed Deere, his contemporary at IH was Fowler McCormick, a descendant of IH founder Cyrus McCormick as well as John D. Rockefeller, the founder of Standard Oil and one of the richest and most successful businessmen in America. Fowler's leadership on social justice issues

WP WORK ON MODEL 60

These images from the Henry Dreyfuss Archives show work done to figure out the contortions required to operate a machine. They were labeled "WP Work on Model 60," and dated July of 1948.
Henry Dreyfuss Archive, Cooper Hewitt, Smithsonian Design Museum

was sound, perhaps reflecting the Rockefeller family ethics. Unfortunately, he and his management team buried IH in debt with ill-fated investments in giant crawlers and home appliances. Meanwhile, Wiman's careful management kept Deere solvent and steady, creating the perfect platform for a vigorous new leader to launch an attack.

In 1955, Wiman passed the torch to his son-in-law, Bill Hewitt. An economics major at UC-Berkeley, Hewitt moved on to Harvard Business School with a small group who would all go on to great success. The circle included Robert McNamara, who would become president of Ford Motor Company and U.S. Secretary of Defense, and Walter Haas Jr., the eventual CEO of Levi Strauss.

The worldly and polished Hewitt proved to be a more aggressive leader than Wiman. One of his first moves was to open the traditionally closed society of Deere management with some outside perspective by inviting the high-profile consulting firm Booz Allen Hamilton to dig deep and offer suggestions for improvements.

In the fall of 1955, he announced his intention to move the company into the agricultural industry's top position. He took the sales and marketing divisions to task as too passive. He also suggested that the engineering departments were too insular and independent, and stated that the marketing department should have a say in design decisions. Along with this, Hewitt called for better interdepartmental communication and mandated more business school education for executives.

With Booz Allen Hamilton's input, Hewitt subtly restructured the company. The old concept of near-complete divisional autonomy was replaced with more centralized management control. The consultant's report suggested the advisory committee, which was mostly a low-stress repository for aging executives, be replaced with one staffed by full-time members from the company's three key divisions: production, sales, and finance. The new committee would be given tremendous knowledge of the company and offer advice to the CEO and other senior executives. The role intended for the committee was fulfilled by the company's longtime senior legal advisor, Edmund Cook, who brought years of experience with the company to his new position.

Hewitt's moves accomplished the nearly impossible goal of decentralizing the design process without destroying the product. At companies, such as General Motors, in the 1980s and 1990s, design by committee met with disastrous consequences. For John Deere, however, it would prove to be an efficient tack.

THE 10 SERIES

By Wayne G. Broehl Jr., from *John Deere's Company*

The insatiable demand for agricultural equipment at the end of World War II momentarily gave the manufacturers a booming "seller's market," in which existing models, however obsolete, were snapped up by avid buyers. But fierce competition soon encouraged innovation and change, and the ten-year period from 1945 to 1955 became noteworthy for product improvement. Deere remained in the forefront in yearly, shorter-term product development, but found itself in an increasing dilemma about one crucial longer-term question: How long should Deere stay with the two-cylinder tractor?

During the 1930s the two-cylinder engine uniquely fulfilled some important demands—for low fuel cost, for simple cooling systems, for straight camshafts with no belts or pulleys, and, in general, for an easily understood, farmer-repairable piece of machinery. Distillates could be used as fuel, the "all fuel" concept at that time giving lower costs. Deere had been successful in selling this concept on its fundamentals—that the two-cylinder engine was as good, or better, on its merits than competitor machines for a wide range of farming conditions. For more than two decades company salesmen had vigorously defended the "Popping Johnnies" against competitor sneers; almost yearly, some trade rival would float rumors that "Deere is going to junk the two-cylinders," hoping to panic buyers away from a model threatened by obsolescence. It became an article of faith among Deere marketing men to fight across the board any attack on the two-cylinder concept. The defense worked very well indeed, for the Models D, A, and B continued to sell in large volume.

After World War II the situation began to change. Many thousands of young farm men had gone off to war to drive tanks, jeeps, weapons carriers, and countless other vehicles, almost all of which were either four- or six-cylinder. When they returned to the farm, often to take over from their weary fathers, would they be willing to stay with their two-cylinder "John Deeres"? Further, some of the most potent substantive arguments for the "two" had lost their credence. Distillate fuels, now relatively higher in price, no longer carried a significant cost advantage. Wartime advances in the catalytic cracking process had made heavy fuels sometimes even more expensive than gasoline; moreover, the heavy fuels required lower compression and, thus, less horsepower. Further, the farmer was demanding larger tractors with higher horsepower, and there were upper limits to the size of two-cylinder engines that could be mounted on a tractor. The horizontal-bore configuration and the diameter needed for such power would soon force so large a size that the tractor could not fit on the row of crops it was to work. The slower speed and lower compression of the distillate-burning engine was an advantage in terms of wear, but a disadvantage in gaining higher horsepower. For this reason, right after World War II, the Waterloo tractors were made available in both a distillate "all fuel" version and a higher-compression gasoline-burning model.

Thus, the Deere dilemma. On the one hand, farmers continued to press for refinements in existing tractors. They wanted better steering control, an independent power takeoff, increased hydraulic power for operating integral equipment (particularly hay balers and forage harvesters), more fuel efficiency, added horsepower for heavier jobs, and, at the same time, more flexible shift-up, throttle-back, fuel-efficient ways of doing lighter work. All of these demanded immediate attention. Competitors were making changes, and if one wanted to stay alive in the marketplace, improvements in existing models had to be constantly provided. This always commanded great amounts of engineering and product planning time.

Yet adaptations to meet competition in the short run always borrowed from longer-term product development. If Deere decided to move to a four-cylinder or six-cylinder tractor, or to both, a much longer

1956 MODEL 50

The replacement for the Model B was the Model 50, and it and the Model 60 featured live hydraulics as well as options for a live power take-off and power steering. The new model also featured what Deere called a "duplex" carburetor, which means it had two intake tracts and throttle butterflies. This is more commonly referred to as a "two-barrel" carburetor.
Mecum Auctions

1953 MODEL 50W

The Model 50 was available in this wide-front configuration, as well as hi-crop versions.
Mecum Auctions

commitment would be required. Deere manufacturing men had often espoused the rule of thumb that a new model of an older machine generally required about 33 percent new parts; a new line incorporating a major engine change would require new designs for 95 percent of the parts.

The Deere management at both of the tractor factories—Waterloo and Dubuque—now faced hard choices. A turn in one direction might foreclose all other opportunities. Perhaps the two-cylinder concept could no longer be sold alone on the fundamental argument that two cylinders per se were as good or better than four or six. Rather, the particular values of the Deere versions needed to be stressed—they were

so well made, so dependable, and so easy to repair and carried such an excellent reputation with the farmer. On this basis it might be possible to sell those same tractors over many future years; the continued success of the Models A, B, and D held considerable promise for this scenario. There would come a point where this could no longer be the case, but had this point been reached?

With the change in the Model D, the question now came of what to do about Models A and B. These enormously popular tractors had been in the line since the mid-1930s; by 1952, there were more than 600,000 of these two Deere models in operation. But, despite their success, they could not continue to

1955 MODEL 60 LPG STANDARD

The Model 60 replaced the Model A for the 1952 season. The LP gas versions became available in mid-1953 as 1954 models. For a time, LP gas was cheaper than the other fuel alternatives. *Keller Collection / Lee Klancher*

compete forever without change. Once again there was the nagging dilemma—the changes could either be minor (cosmetic or small refinements), or major, or, more basically, they could entail scrapping old models and introducing tractors with more cylinders.

By this time there were strong advocates, including Charles Wiman, for phasing out the two-cylinder tractors. Dubuque engineers had already developed an upright two-cylinder engine for the Model M, and they were experimenting with another upright multiple-cylinder engine for self-propelled harvesters.

Duke Rowland, who still headed all of the company's tractor production, was adamantly reluctant to scrap Waterloo's very successful tractors, however. They were selling well in the field and his organization

had its manufacturing on so efficient a basis that the factory was the biggest profit-making unit in the system. Over and over, he told the management in Moline, "We would be run over by our competition if we stopped now to change—we couldn't make the shift within nine months of downtime—we'll go broke, we'd be just like Ford when it changed from the Model T to the A." These arguments worried Wiman, for they called up images of the Great Depression that he had faced early in his career; still, though Wiman was genuinely ambivalent on the timing of the change, he nevertheless advocated making it earlier rather than later.

Once again the authority vested in the factories by the policy of "decentralization" intruded to decide

1954 MODEL 60 LPG HI-CROP

The hi-crop version of the 60 offered 32 inches of ground clearance. Only 212 hi-crops were built, with 15 of them being LP gas versions. *Mecum Auctions* the question. Rowland's conservative posture held sway, and the decision was made to upgrade Models A and B, but to retain the two-cylinder motor. In 1952, two new models were introduced—the Model 50, to replace the B, and the Model 60, to replace the A. Both had many innovations. Each had duplex carburetion (a separate carburetor for each cylinder), which allowed more precise fuel metering and helped increase horsepower (the Model 50 was rated at 20.62 drawbar horsepower and 26.32 belt horsepower). Both were row-crop tractors, like their predecessors, and the 60 was also available in a "hi-crop" model of particular value to producers of tall, bushy, bedded crops like sugarcane, flowers, etc. These two models were also the first of the Deere tractors available with "live" power shafts, which provided continuous power for operating equipment driven by a power takeoff (PTO). A continuous running or an independent PTO became a necessity for the postwar versions of hay balers and field forage harvesters, for both machines contemplated stopping the forward travel of the tractor from time to time while continuing the PTO for the baler or forage harvester.

An interesting issue surfaced at this time, that of interchangeability. Farmers wanted to be able to use any given power equipment, whatever the make, in whatever combination desired; thus, the lack of compatibility of PTOs became an industry-wide problem. Once again, as the industry had done just after World War II with hydraulic control standardization, the Farm Equipment Industry Engineering Advisory Committee set up an industry-wide subcommittee on the PTO to standardize such things as hub and spline dimensions, drawbar hitch and power takeoff location, and drawbar vertical load dimensions. Out of this came a set of standards in 1958 that were uniformly adopted by the industry. Each manufacturer was free to develop his own engineering of the internal mechanisms that would lead to the external PTO, but the couplings themselves would all be standard.

In 1953, the Model 70 arrived in the product line. Originally, the model was available with gasoline,

"all-fuel," or LP gasoline options but soon was offered with a diesel option, and thus it became the first Deere diesel row-crop tractor. In its Nebraska tractor test, the Model 70 diesel set a new industry fuel economy record, bettering all previously tested row-crop tractors of all manufacturers. By 1954, Deere engineers had perfected another "industry first," an optional factory-installed power-steering system for all three of the new models. It utilized built-in hydraulics to control the steering mechanism, differing from so-called "add-on" systems that used externally mounted motors on the steering shafts or hydraulic cylinders hooked up to tie rods. Deere's power steering was a breakthrough for the industry, soon widely adopted by other manufacturers. Again, though, the engineering time taken to develop it was at the expense of an upright, multiple-cylinder engine.

All of these models were manufactured at Waterloo, with all of the engineering done there. Meanwhile, the Dubuque engineers were equally busy. In 1953 they brought out a new Model 40 in both a standard tractor, a tricycle, and a crawler, replacing the Dubuque-manufactured Models M, MT, and MC. In effect, the Model 40 series was a smaller counterpart of the Model 50, 60, and 70 tractors built in Waterloo. (The 40T model was rated at 17.16 drawbar horsepower and 21.45 belt horsepower; the crawler was rated at 15.11 drawbar horsepower and 21.45 belt horsepower). In 1955, six years after the introduction of the historic Model R diesel, its own replacement arrived, the Model 80 diesel. This was a gargantuan machine, weighing some 7,850 pounds, and it was quite powerful—57.49 horsepower at the belt, 46.32 at the drawbar. It incorporated many of the most popular features on the other numbered models, such as optional power steering and "live" power shaft, and also included a six-speed transmission. The 80 was capable of pulling a twenty-one-foot disk and working up to 126 acres with it in a single day. A large 32.5-gallon fuel tank allowed the operator to stay in the field many hours before making a refueling stop.

1955 MODEL 70 STANDARD LPG

The Model 70 replaced the Model G with a bang, offering the farmers of the mid-1950s a popular combination of horsepower and versatility. It also broke the Model R's long-standing fuel economy record. *Mecum Auctions*

1956 MODEL 70 HI-CROP DIESEL

Introduced in 1954—along with the Model 80—the diesel version of the Model 70 was started with a V-4 auxiliary gasoline engine. *Mecum Auctions*

1954 MODEL 40H FIRST PRODUCTION

Designed to replace the Model M, the 40 was a direct competitor to the incredibly successful Ford Model 9N. In 1955, a Model 40 Hi-Crop cost $1700. Only 294 Model 40 Hi-Crops were built. *Keller Collection / Lee Klancher*

**1954 MODEL 40V
FIRST PRODUCTION**

The Model 40 was available in seven different configurations, including a hi-crop, utility, a crawler, wide and narrow front, and this unusual "Special" configuration built for sweet potato, peanut, and cotton growers. *Keller Collection / Lee Klancher*

The new models—the 40-50-60-70-80 group—were well received, particularly for the numbers of refinements and innovative new ideas they incorporated. But the handwriting was on the wall about the staying power of any two-cylinder tractor, no matter how good. Waterloo knew this as well as anyone. By this time, Duke Rowland had left his position heading tractor production and Maurice Fraher assumed his post. Both Fraher and Gust Olson, the Waterloo Works

manager, were less resistant to change. So a decision was taken, sometime in early 1953. (The exact date is no longer determinable, the whole operation being so secret that no records were kept!) A select group of Waterloo engineers and design men were to be pulled off their regular duties and set to work on the task of developing an upright, multiple-cylinder motor. Their charge: "Start with a blank drawing board."

MODEL 40C

The Model 40C was introduced with a three-roller track, and later versions, like this one, had a choice of four (as shown) or five rollers. *Mecum Auctions*

A turning point in John Deere history was set in August 1946 when Ed Osgood forgot to make hotel reservations for a wedding in Santa Barbara.

Osgood invited his friend, the thirty-two-year-old William Hewitt, to come down to Santa Barbara to hang out for a few days before the wedding. Hewitt met his

friend at the Coral Club, a private tennis club, and found him playing a game against a striking redhead, Patricia Deere Wiman, better known to all as "Tish." When Hewitt asked Osgood where they would stay that night, his friend sheepishly admitted he had not reserved a room.

Hewitt stated that Tish piped up that they could stay at her family's guest house, according to Hewitt's account of the meeting in *The Hewitt Family History*. The Wiman family cottage would house Hewitt, and he and Tish hit it off. Less than a year later, Bill asked Tish to marry him.

Tish's father was Charles Wiman, a president at Deere, and she was the great-great-granddaughter of the company founder, John Deere. Athletic, vivacious, and worldly, Tish had a lifetime love of horse shows and competition, fashion, and philanthropy.

Hewitt was a territory manager for the Pacific Tractor and Implement Company at the time, and would of course move on to work for Deere. He moved to Deere in 1948, was named to the Board of Directors in 1950, and was appointed CEO in May 1955. He would oversee perhaps the most significant new model introduction in company history, the release of the New Generation, and was widely lauded for his work growing the company.

He and Tish were a striking couple, with lots of visibility in the society pages as they sailed through life. The tractor on these pages was used to perform maintenance on Tish's parent's winter home in Santa Barbara.

If Bill's friend had made those hotel reservations, this beautifully worn old tractor—and more significantly, an important chapter in Deere history—may have been written very differently.

1953 MODEL 40S

The tractor photographed here was owned by Charles Deere Wiman, who was the father of Pattie "Tish" Deere Wiman, the great-great-granddaughter of John Deere. Tish married eventual Deere CEO William Hewitt. *Keller Collection / Lee Klancher*

DEERE-WIMAN FAMILY TRACTOR

The historic Deere was sold in 1953 to Mr. and Mrs. Colonel Charles Deere Wiman. The couple bought the tractor to help with maintenance of their winter home in Santa Barbara. The Model 40 was continuously maintained by Goleta Tractor Service in Santa Barbara until 1982, when it was traded in for a John Deere 112 riding tractor.

TWO-CYLINDER 10 SERIES DATA

Model	Type	Model Years	Notes	HP	Nebraska Test #
50	Row-Crop	1952–1956	Replaced Model B. First built 07/24/1952. Last built 05/14/1956. All-fuel, gas, or LP engine.	33	486 (1952 gas), 507 (1953 all fuels), 540 (1955 LP)
60	Row-Crop	1952–1956	First built 03/12/1952. Last built 05/18/1956. All-fuel, gas, or LP engine.	46	490 (all fuels)
60S	Row-Crop (standard)	1952–1956	All-fuel, gas, or LP engine.	-	-
60 (regular)	Row-Crop	1952–1956	All-fuel, gas, or LP engine.	46	472 (1952), 513 (1953 LP)
60-O	Orchard	1952–1956	All-fuel, gas, or LP engine.	-	-
60H	Hi-Crop	1952–1956	All-fuel, gas, or LP engine.	-	-
60S6 (High-seat)	Row-Crop	1955–1956	All-fuel, gas, or LP engine.	-	-
70	Row-Crop	1953–1956	All-fuel, gas, or LP engine.	48	493 (gas), 506 (all fuels), 514 (1953 LP)
70 Hi-Crop	Row-Crop	1953–1956	All-fuel, gas, or LP engine.	-	-
70S	Row-Crop (standard)	1953–1956	All-fuel, gas, or LP engine.	-	-
70D	Row-Crop	1954–1956	Diesel engine.	-	528
40S	Row-Crop (standard)	1953–1955	Replaced Model M. First production 01/09/1953. Last 10/17/1955. All-fuel or gas engine.	24	504 (gas), 546 (all fuels)
40C	Agricultural Crawler	1953–1955	First production 11/04/1952; last 10/17/1955. All-fuel or gas engine.	-	505
40T	Row-Crop (tricycle)	1953–1955	All 40s built at Dubuque Tractor Works. All-fuel or gas engine.	-	-
40H	Hi-Crop	1953–1955	All-fuel or gas engine.	-	-
40W	Two-Row Utility	1953–1955	All-fuel or gas engine.	-	-
40V	Special	1953–1955	All-fuel or gas engine.	-	-
40U	Utility	1953–1955	All-fuel or gas engine.	29	-
80	Row-Crop	1955–1956	Diesel engine (gas start).	65	567

MODEL 80

The Model 80 was the last of the series to be transformed from a letter designation to a number, and it was introduced in mid-1955. The change turned out to be worth the wait, as the 80 was a significant improvement of the excellent Model R it replaced.

The engine was updated and offered a nice horsepower increase that boosted the rating from forty-eight to sixty-five. The starting motor was improved, and the five-speed transmission of the Model R was replaced with a six-speed on the 80. The Model 80's PTO offered better power transfer, the hydraulics were upgraded, the rear wheel tread could be set wider, and fuel capacity bumped up a bit.

The big machine's optional power steering was a game-changer, as it made piloting the heavy machine a relative piece of cake. By late 1956, the Model 80 would be replaced with a newer model, so this short-lived model only saw about 3,500 units produced.

1956 MODEL 80

The Model 80 looks a lot like the Model R it replaced, but there are significant improvements under the skin. This one is equipped with optional cane and rice field equipment, which was an optional package and production numbers are not known. *Mecum Auctions*

THE 20 SERIES

Color was all the rage in the 1950s. Eastman Kodak introduced 35mm color film in 1950. Flashy two-color paint schemes emerged on Oldsmobiles and Buicks in 1954, and the rest of the American auto industry followed suit in 1955. At General Motors, designer Harley Earl oversaw a seventy-five-member Styling Color and Interior Design Studio.

The concept of planned obsolescence drove Detroit's fixation with color and style. By making next year's model more stylish, you created demand. Shallow improvements such as new fabrics and bigger tail fins were easy ways to sell machines.

Henry Dreyfuss was not swayed. "The realistic manufacturer is working towards a fundamental improvement in his product . . . that will give the consumer a really convincing reason for trading in the old model," he wrote.

His firm's work with the 20 series included improving comfort and ease of operation as well as adding subtle design cues. His most visible contribution to the 20 series, however, was a splash of carefully chosen yellow.

NEW LOGO

The John Deere logo was redesigned in 1956 to a simpler, more elegant design that signified the company's growing confidence and brand recognition. *Keller Collection / Lee Klancher*

1958 MODEL 320

The smallest of the 20 series two-cylinder tractors was built in Dubuque, and shared parts with the Model 40, M, and 420. Early models had vertically oriented steering wheels, and late 1958 models had the "slant steer" feature shown here. This tractor spent its working life on a California blueberry farm and was restored by Henry Delbridge. *Mecum Auctions*

1956 MODEL 420H ALL-GREEN VARIANTS

The 420 was released in November 1955, before the rest of the 20 series tractors. The 1956 models are easy to identify because they are all green, rather than having the yellow inset panels. *Mecum Auctions*

MODEL 420 LPG

When the rest of the 20 series was introduced in summer 1956, the 1957 and later 420s received the two-tone paint job like the rest of the series. *Watral Collection / Lee Klancher*

1958 MODEL 420S

Most of the 1958 model 420s received a slanted steering wheel, as shown here. *Mecum Auctions*

1956 MODEL 420U

The 420 came in a dizzying number of variants, including utility models. The 420 industrial was replaced in 1958 by the 440, which was the industrial version of the 435. A handful of Holt forklift variants were built as well (*Green* magazine cites less than twenty-six forklift versions built). *Mecum Auctions*

1957 MODEL 420C

The crawler variant of the 420 was offered in both agricultural and industrial versions, and was available with four- or five-roller tracks. *Mecum Auctions*

1957 MODEL 520

The replacement of the Model 50—and prior to that, the Model B—the 520 had few variants and did not see changes to paint color of the dash / steering angle of other 20 series two-cylinder models. *Kline Collection / Lee Klancher*

"The whole darn thing looks just too busy—it looks like a Persian carpet."

— Bill Hewitt describing Lanz tractors in a late-1950s letter to Henry Dreyfuss

EARLY LANZ TRACTOR

Heinrich Lanz AG was founded in Germany in 1859 and was producing award-winning agricultural machinery by 1866. This large Lanz tractor is at work in the early 1920s. *Library of Congress / Harris & Ewing Photographic Collection*

In 1859, Heinrich Lanz founded a company importing American agricultural equipment to Germany. The company moved on to manufacturing equipment and steam-powered engines, and by 1902 was the largest agricultural company in Germany, employing more than one thousand workers.

The company's best-known tractor, the Bulldog, was developed by Lanz employee Fritz Huber and introduced in 1921. The simplicity of the tractor's two-stroke, hot-bulb engine and its ability to burn nearly any kind of low-grade fuel was key to the Bulldog's longevity.

The machine was produced until 1960, with more than 200,000 units sold.

During World War II, most of Lanz's Mannheim plant was bombed out of existence. After the war, a new, smaller tractor, known as the Alldog, was rushed into production. Using an engine from a substandard supplier, it was unreliable and caused serious damage to the Lanz reputation.

The Lanz and Deere companies had discussed a purchase since the early 1950s, and by 1956 Deere was increasingly interested in expanding its presence in world markets. Purchasing a German company would give Deere immediate foot in that country. On the surface at least, the deal made sense.

The purchase was completed in 1956. Once they installed new management and began to dig into the

LANZ BULLDOG

The cornerstone of the Lanz tractor lineup was the Lanz Bulldog tractor, the earliest version of which appeared in the 1920s. The Lanz Bulldog is one of the most prolific models in history. More than 200,000 units were built from 1921 to 1960, and all used this incredibly simple single-cylinder, two-stroke engine design. This Bulldog was photographed on September 15, 1938. *Interfoto*

details of their new branch, Deere officials discovered the German company was in serious trouble. The tractor product line was too broad and the engines they used were either outmoded or substandard. Company records were poorly kept, and what records there were showed a vast backlog of customer complaints. Deere quickly decided to create a new line using the small utility engines and design from its Dubuque plant. But the new tractors were not built ruggedly enough for European demands, nor did they have the necessary transport road speed.

Beyond the problems of engineering for a foreign market, Deere & Company managers also had serious difficulty adapting to the German culture. The German engineers appeared hidebound and conservative to American management. The fact that German line workers were accustomed to freshly tapped glasses of beer at break time made American management nervous. When the policy was changed so that only small bottles of beer were available, a strike ensued just as the new models were being built. The keg beer was reinstated and remained available on the factory floor for more than twenty years.

The Lanz experiment was successful in that it gave Deere & Company a foothold in Europe, but the early results were not what Deere officials were content to accept.

BEET HARVEST

By the time the photo of this Bulldog was taken in the 1950s, the Bulldog was getting long in the tooth. Deere purchased Lanz in 1956 to gain a foothold in the European market. *Interfoto*

1960 JOHN DEERE-LANZ MODEL 3016

The John Deere-Lanz Model 3016 is a version of the Lanz Bulldog that was built after Deere purchased the German tractor maker in 1956. One of the first changes that Henry Dreyfuss implemented was to change the Lanz paint scheme from blue to green. *Keller Collection / Lee Klancher*

1960 JOHN DEERE-LANZ MODEL 3016

The Lanz Bulldog used a two-stroke, hot-bulb diesel engine, a design that features no valves or camshaft and uses heat and compression to ignite the air-fuel mixture. The mixture is sprayed through ports in the cylinder wall into the cylinder, where it contacts the red-hot portion of the cylinder head. Starting hot-bulb engines requires the bulb to be heated, which can take several minutes. The engines often started on gasoline and switched over to a low-cost fuel once warm. *Keller Collection / Lee Klancher*

1960 JOHN DEERE-LANZ MODEL 500

The John Deere-Lanz Model 300 and 500 were introduced in the early 1960s. They were handsome, functional machines, but were expensive and somewhat underpowered compared to the competitive tractors built by Massey-Ferguson and Ford. *Interfoto*

GERMAN-BUILT JOHN DEERE MODELS 1960–2020

Model	Production Dates	Model	Production Dates	Model	Production Dates	Model	Production Dates	Model	Production Dates
300 Lanz	1960–1966	2840	1977–1979	2955	1987–1992	6215	2003–2007	6150M	2013–2015
500 Lanz	1960–1966	4040	1977–1984	1550	1987–1994	6415	2003–2007	6170M	2013–2015
100 Lanz	1962–1966	4240	1977–1984	1750	1987–1994	6615	2003–2007	6175R	2014–
700 Lanz	1962–1966	4440	1977–1984	2250	1987–1994	6715	2003–2007	6195R	2014–
1020 (Also Dubuque, Iowa & Mexico)	1965–1973	3030	1978–1979	2450	1987–1994	5620	2003–2008	6215R	2014–
310	1966–1967	1840	1979–1982	2650	1987–1994	5720	2003–2008	6110M	2015–
510	1966–1967	1640	1979–1987	3155	1988–1992	5820	2003–2008	6110R	2015–
510	1966–1967	2040	1979–1987	1950	1988–1994	7130	2007–2011	6120M	2015–
710	1966–1967	2940	1980–1982	3255	1991–1993	7430 Premium	2007–2011	6120R	2015–
200 Lanz	1966–1968	940	1980–1986	3055	1992–1993	7530 Premium	2007–2011	6130M	2015–
200	1966–1976	1040	1980–1987	6100	1992–1997	6230	2007–2012	6130R	2015–
2020	1967–1972	1140	1980–1987	6200	1992–1997	6330	2007–2012	6145M	2015–
920	1967–1975	2140	1980–1987	6300	1992–1997	6430	2007–2012	6155M	2015–
1120	1967–1975	3040	1980–1987	6300	1992–1997	7230	2007–2017	6175M	2015–
1520	1968–1972	3140	1980–1987	6400	1992–1997	7330	2007–2017	6195M	2015–
820	1968–1973	4040S	1981–1984	6400	1992–1998	5090R	2009–	6230R	2016–
2120	1969–1972	4240S	1981–1984	6600	1993–1997	5100R	2009–	6250R	2016–
3120	1969–1972	2040S	1981–1987	6800	1993–1997	6534	2009–2011	6140M	2020–
1530	1973–1975	2350	1983–1986	6200	1993–1998	5080R	2009–2016	6225	unknown
2130	1973–1977	2550	1983–1986	6900	1994–1997	6170R	2011–2014	6325	unknown
1030	1973–1979	2750	1983–1986	6500L	1994–1998	6190R	2011–2014	6425	unknown
1630	1973–1979	2255	1983–1987	6506	1995–1997	6210R	2011–2014	6525	unknown
1830	1973–1979	2950	1983–1988	6405	1998–2002	6105R	2012–2014		
3130	1973–1979	3640	1984–1987	6605	1998–2002	6115R	2012–2014		
830	1974–1979	3150	1985–1987	6110	1999–2002	6125R	2012–2014		
930	1974–1979	4350	1985–1993	6210	1999–2002	6105R	2012–2016		
1130	1974–1979	3050	1986–1993	6310	1999–2002	6115R	2012–2016		
2040	1976–1982	3350	1986–1993	6410	1999–2002	6125R	2012–2016		
2240	1976–1982	3650	1986–1993	6510L	1999–2002	6130R	2012–2016		
2440	1976–1982	1350	1986–1994	6420	2002–2006	6140R	2012–2016		
840	1976–1986	1850	1986–1994	6520L	2002–2006	6150R	2012–2016		
		2850	1986–1994	6120	2002–2007	6105M	2013–2015		
		2155	1987–1992	6220	2002–2007	6115M	2013–2015		
		2355	1987–1992	6320	2002–2007	6125M	2013–2015		
		2555	1987–1992	6215	2003–2005	6140M	2013–2015		
		2755	1987–1992	6515	2003–2005				

MODEL 620

1956 MODEL 620 STANDARD LP

Shipped to Florida and used in orchards, this Model 620 tractor is equipped with a very rare special-order, under-slung exhaust that exits at the rear of the machine.

1956 MODEL 620 STANDARD LP

Only thirty-seven Model 620 Standard LPs were built. Note that production of the 620 Orchard lasted until 1960, after the rest of the 20 series was replaced by 30 series two-cylinder models. *Keller Collection / Lee Klancher*

1958 MODEL 620

The 620 wide-front standard model pictured here is equipped with power steering. Some 620s were assembled at a Deere plant in Monterrey, Mexico, and sold in Mexico. Those models had "John Deere de Mexico" on their serial number plates. *Mecum Auctions*

1957 MODEL 720

One of the most efficient
and refined two-cylinder
John Deere tractors,
the Model 720 remained
popular and in use on
farms for decades.
*Glass Collection /
Lee Klancher*

**1957 MODEL 720
HI-CROP**

This rare variant is the first
of five gasoline-engine,
hi-crop Model 720s built.
This tractor was part of the
Jim Mills collection.
Mecum Auctions

1957 MODEL 720 STANDARD

Most of the 720 standards were sold with diesel engines. This is another gasoline-engine version, of which 520 were built in 1957 and 1958.

Renner Collection / Lee Klancher

1957 MODEL 720

The diesel engine in the Model 720 was one of the most fuel-efficient tractor engines of the era. The diesel engine was started with either an auxiliary gas motor (shown here) or an optional electric starter (available later in 1958). The 1958 engine was improved with a new cylinder head, crankshaft, and more.

Vinopal Collection / Lee Klancher

1957 MODEL 820

In addition to the two-tone paint job, major improvements from the Model 80 were larger brakes, a new seat, and some new options. Optional cabs were offered for the 720 and 820 models. *Mecum Auctions*

TWO-CYLINDER 20 SERIES DATA

Model	Type	Model Years	Notes	HP	Nebraska Test #
320	Utility	1956–1958	Gas engine.	21.5	-
320U	Utility	1956–1958	Gas engine.	21.5	-
320S (standard front)	Utility	1956–1958	Gas engine.	21.5	-
320V (Southern special)	Utility	1956–1958	Gas engine.	21.5	-
420C	Agricultural Crawler	1956–1958	Gas or all-fuel engine.	-	601
420H	Hi-Crop	1956–1958		-	
420I	Special Utility	1956–1958	Gas, LP, or all-fuel engine.	-	
420S	Row-Crop	1956–1958	All-fuel engine.	23	600
420T-N	Row-Crop	1956–1958	Single front, tricycle.	-	-
420T-RC	Row-Crop	1956–1958	Dual front, tricycle, 21-inch clearance.	-	-
420T-W	Row-Crop	1956–1958	Adjustable wide front.	-	-
420U	Utility	1956–1958	Forklift, Model 3T, Holt Mfg. built 25. Gas, LP, or all-fuel engine.	-	-
420V	Special	1956–1958	Semi Hi-Crop.	-	-
420W	Two-Row Utility	1956–1958	Low adjustable wide front, 20.5-inch clearance; 8 were built. Gas, LP, or all-fuel engine.	28	599
520	Row-Crop	1956–1958	Gas, LP, or all-fuel engine.	37	590 (LP), 592 (all fuels), 597 (gas)
620	Row-Crop	1956–1958	First 620 built 06/07/1956. Last 620 built 07/25/1958. Gas, LP, or all-fuel engine.	49	591 (LP), 598 (gas), 604 (all fuels)
620O	Orchard	1956–1958	Last 620O built 02/01/1960 (into 630 production). Gas, LP, or all-fuel engine.	-	-
620HC	Row-Crop (Hi-Crop)	1956–1958	Gas, LP, or all-fuel engine.	-	-
620S	Row-Crop (standard)	1956–1958	Gas, LP, or all-fuel engine.	-	-
720	Row-Crop	1956–1958	Standard and Hi-Crop. Gas, LP, all-fuel, or diesel engine (gas start; electric start 1958).	57	593 (LP), 594 (diesel), 605 (gas), 606 (all fuels)
820	Row-Crop	1956–1958	820 also built in Germany 1968–1973 (used same model #). Diesel (gas start).	73	632 (see 830)

THE 30 SERIES

While the introduction of the 30 series was hardly a ground-breaking update—the mechanical improvements were clearly refinement rather than raw innovation—the line represents a simple technology elevated to high art. Deere had refined the two-cylinder engine concept taken from the Waterloo Boy tractor into a highly evolved line of efficient farm machines.

The engines, particularly the two-cylinder diesel, offered class-leading fuel economy and tremendous longevity and legendary reliability. The hydraulic and hitch technology was state-of-the-art, and operator comfort as well as ease of operation were also class leaders.

Interestingly, a portion of the Deere excellence in this era can be traced back to the relationship with Henry Dreyfuss. While all the major agricultural manufacturers had relationships with industrial designers, Dreyfuss was a special talent.

The cover of the May 1, 1951, *Forbes* magazine shows Dreyfuss in his brown suit, surrounded by his best-known creations, including the *20th Century Limited*, a Hoover vacuum cleaner, and the Big Ben alarm clock.

Those designs defined Dreyfuss as one of the best industrial designers of the time. Dreyfuss, along with Raymond Loewy, Walter Dorwin Teague, and Dreyfuss' former instructor, Norman Bel Geddes, brought style and elegance to daily American life. They also made some of America's largest companies more profitable.

All had talented staffs to execute their visions, but Dreyfuss took a much more hands-on approach than the others. Dreyfuss was not involved much with sales and production, but he was heavily involved in the design and engineering of every product. He personally approved every design generated by his firm, and he would drop in at any hour to give feedback to his people.

In *The Man in the Brown Suit*, author Russell Flinchum recounts a story told to him by HDA designer Jim Conner, who was working on the roll bar

HENRY DREYFUSS

Industrial design played a key role in making John Deere tractors exceptional, and Dreyfuss was one of the finest of the era. He would take only fifteen clients, and deliberately kept his company small and personal. He learned to drive a tractor to understand them better and kept a John Deere at his home in Pasadena. His contributions brought an elegance to the two-cylinder John Deere machines and would go deeper with more contemporary lines.

Henry Dreyfuss Archive, Cooper Hewitt, Smithsonian Design Museum

for John Deere tractors. Dreyfuss suggested the drawings were too irregular and bristled when Conner recommended they run this by Bill Purcell, another HDA designer. Dreyfuss was incredulous that his authority on design would be questioned.

Dreyfuss kept his client list short so that he could oversee each account personally. He maintained offices in Pasadena, California, and New York City, allowing plenty of face time with clients. A master of quick sketches, Dreyfuss could execute his renderings upside down when meeting with a client seated across a table.

"Our goal has always been to become a member of the client's 'family', remaining in touch with his problems, co-operating closely on his current merchandise, but also keeping a sharp eye out for future programs," Dreyfuss wrote in *Designing for People*. "We feel that we must be faces and personalities—not merely a voice on the telephone, the signature on a letter, or an initial on a drawing."

The Dreyfuss approach was effective on a number of levels, including compensation. He charged $150 per hour for his services, a figure that shocked even his competitors.

Deere CEO Bill Hewitt embraced Dreyfuss's methods. Due to Hewitt's policy of opening the company's previously closed environment, Dreyfuss was able to work more closely than ever with Deere's engineering and design staff during the creation of the New Generation tractors.

The 30 series of two-cylinder tractors was an outstanding line of machines, but the stroke of genius that would carry Deere to the top of the heap was yet to come.

HUMAN FACTORS ENGINEERING

Dreyfuss was a pioneer in the concept of fitting the machine to the body. His book, *Designing for People*, appeared in 1955 and included detailed human ranges of measurements to allow designers to create machines to fit people rather than vice versa. *Henry Dreyfuss Archive, Cooper Hewitt, Smithsonian Design Museum*

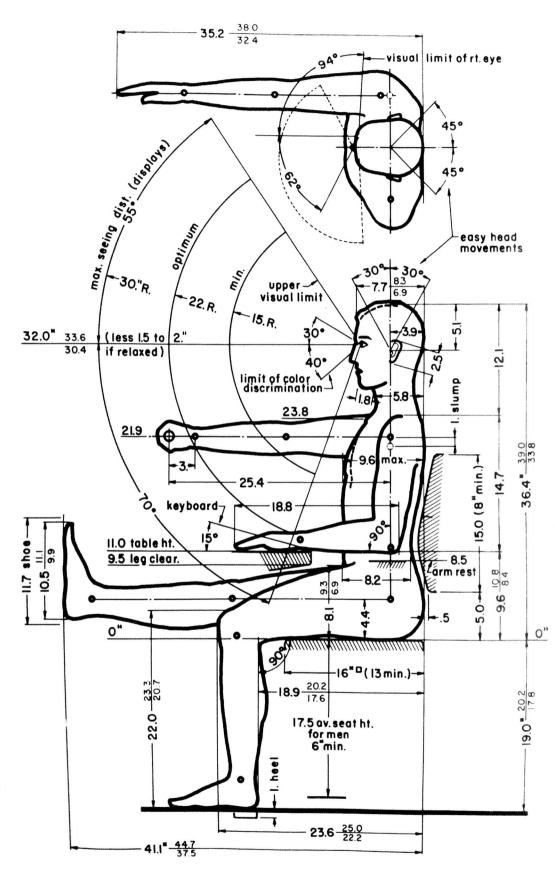

1959 MODEL 430W LP

The Model 430 was built between 1958 and 1960. Total production was 14,697. The LP version is a rare machine, with only sixty-eight built.

Keller Collection / Lee Klancher

1959 MODEL 330

Featuring new styling and minor mechanical upgrades, the 30 series gave John Deere something "new" to sell while the company developed an innovative new line.
Mecum Auctions

MODEL 530

The 530 featured push-button starting, revised controls, and positioning, and was only offered as a row-crop tractor with dual or single front wheels or a wide front option. It was the smallest tractor built at the Waterloo factory.
Henry Dreyfuss Archive, Cooper Hewitt, Smithsonian Design Museum

1959 MODEL 630 HI-CROP

The Model 630 was available in general purpose, standard tread, and hi-crop versions. The general purpose had four front ends to choose from: a single front wheel, wide-front, dual narrow front, and the dual Roll-O-Matic. *Keller Collection / Lee Klancher*

LAST PRODUCTION 630 HI-CROP

The 630 was produced from 1958 to 1960. Around 18,000 examples were built. This 1959 tractor bears serial number 6315987, making it the last 630 hi-crop produced. *Keller Collection / Lee Klancher*

MODEL 730 STANDARD DIESEL

The look was given some crisp new touches and operator comfort was improved, but the 730 engine and transmission were largely unchanged from the 720.

Purinton Collection / Lee Klancher

**MODEL 730
OPERATOR'S VIEW**
New features on the
730 included a new
instrument panel,
slanted steering wheel,
and improved control
lever locations.
*Purinton Collection /
Lee Klancher*

When Bill Hewitt took over as the leader at John Deere, his goal was to make it the world's top tractor manufacturer. Doing so would require penetration into foreign markets. He admitted to *Forbes* magazine that Deere & Company was behind the curve—competitors had been strong in overseas markets since the dawn of the twentieth century. Deere & Company had no sales offices outside of the United States and Canada.

During the last years of Charles Wiman's management, the company explored a presence in Scotland. Expansion into South America was also considered and rejected. Under Hewitt's direction, the company looked to Mexico and Germany, authorizing an assembly plant in Mexico in 1956, the same year it bought the German company Lanz. A new facility in Rosario, Argentina, was built shortly after that.

As Deere was building in Argentina, Juan Peron encouraged Argentine manufacturers to build their own tractors. The result was the IAME tractor company, which built the Pampa tractor, a virtual copy of the Lanz Bulldog. The machine's simple engine, able to burn almost any fuel—even animal fat—was ideal for the market.

As many as 3,500 blue Pampas were rumored to have been built in the mid-1950s, but the factory lasted only a few years. According to those who have traveled to Argentina in search of John Deere tractors, the Pampa tractors are easy to find. "They are a dime a dozen," one finder named Steve says. Steve has traveled extensively in Argentina and Bolivia, and has been to the archives in Rosario. He finds the machines by wandering.

For a tractor vagabond like Steve, finding desirable machines in strange countries is not overly difficult. The trick is getting them home. "Argentina is full of graft and corruption," he says. "It's not for the meek of heart to bring stuff outside of Argentina." Still, he has managed to ship hundreds of tractors to the United States.

Deere's struggles with Argentina were no less difficult. The government's policies changed constantly, and the company lost money there from the mid-1960s to the early 1970s. Several foreign business leaders were kidnapped and killed by guerillas during the early 1970s, and inflation ran rampant. When terrorist activity abated and the government stabilized in the late 1970s, an agricultural depression hit, and interest rates skyrocketed. When import restrictions were reduced, the flood of lower-cost tractors into the country made it difficult for Deere to compete.

International markets present unique challenges. Some, like Argentina, offer a range of problems that are impossible to control. You just have to go with the flow and hope for the best. It's the South American way.

MODEL 730 HI-CROP

The John Deeres initially built in Rosario, Argentina, were done as assembly rather than true manufacturing, with much of the tractor built at Deere plants in the United States and then shipped to Argentina for final assembly. This Model 730 tractor was one of those, and it was constructed in the United States on August 5, 1958, and shipped, probably in pieces, for assembly at the Deere & Company facility at Rosario, Argentina. When two-cylinder production was halted in the United States, production of the 730 continued in Rosario, Argentina.

ARGENTINE MODEL 730

The Argentine 730s were equipped with diesel engines and electric starts. Four versions were built: standard, row crop (both narrow and wide front), and hi-crop. The last 730 built in Argentina was constructed on August 1, 1971.

ARGENTINE MODEL 730

Argentine Model 730s built after the mid-1960s had nearly every part stamped with "Industria Argentina" or "Industria AG." U.S. Model 730 production ended in July 1970. According to records at the Rosario plant, 33,032 Model 730 tractors were built in Argentina.

MODEL 445

The John Deere 445 is an orchard tractor built in Deere's factory in Rosario, Argentina. The model is similar to the Model 435, both of which are powered by a General Motors 2-53 two-stroke diesel engine.

MODEL 445

This Model 445 came to the Keller collection from an orange grove near Mendoza, Argentina. The machine bears serial number 446003. John Deere also offered tricycle, vineyard, and "economy" versions of the 445 in Argentina. The latter featured smaller wheels and was shipped without fenders.

JOHN DEERE MODELS BUILT IN ARGENTINA			
Model	**Production Years**	**Model**	**Production Years**
730	1961–1971	3530	1975–1980
445	1963–1970	4530	1975–1980
1420	1970–1975	3330	1977–1980
2420	1970–1975	3440	1979–1981
3420	1970–1975	3540	1983–1988
4420	1970–1975	3350	1986–1993
2330	1975–1980	2850	1986–1994
2530	1975–1980	3550	1988–1993
2730	1975–1980		

MODEL 830 DIESEL FIRST PRODUCTION

The Kellers purchased this historic machine from a well-known collector in Iowa named Martha who would tour the collector's circuit on the machine. The deal required months of negotiation, and when the time finally came, Walter and Bruce Keller went to her place in Iowa to pick it up. Walter photographed her and the tractor just before they left, and the photo remains one of his favorites. "This tractor is named 'Martha,'" Walter explains. "It's a tractor she loved."

Keller Collection / Lee Klancher

MODEL 830 DIESEL FIRST PRODUCTION

This tractor is the first production model, serial number 830000, constructed on August 4, 1958.The Model 830 was built until 1960. During that time, 6,712 were produced. This Model 830 is equipped with electric start, but the model also could be ordered with a V-4 starting engine. *Keller Collection / Lee Klancher*

TWO-CYLINDER 30 SERIES DATA

Model	Type	Model Years	Notes	HP	Nebraska Test #
330S (standard)	Row-Crop	1958–1960	Gas engine.	21.5	-
330U	Utility	1958–1960	Gas engine.	-	-
430S	Row-Crop (standard)	1958–1960	Gas, LP, or all-fuel engine.	28	599 (gas), 600 (all fuels)
430T	Row-Crop (tricycle)	1958–1960	Gas, LP, or all-fuel engine.	-	-
430U	Utility	1958–1960	Gas, LP, or all-fuel engine.	-	-
430H	Row-Crop (Hi-Crop)	1958–1960	Gas, LP, or all-fuel engine.	-	-
430V	Row-Crop (special)	1958–1960	Gas, LP, or all-fuel engine.	-	-
430W	Row-Crop (utility)	1958–1960	Gas, LP, or all-fuel engine.	-	-
430C	Crawler	1958–1960	Gas, LP, or all-fuel engine.	-	-
430F				-	-
530	Row-Crop	1958–1960	First production 06/10/1958. Last 02/1960. Gas, LP, or all-fuel engine.	37	590 (LP), 592 (all fuels), 597 (gas)
630S	Row-Crop (standard)	1958–1960	Gas, LP, or all-fuel engine.	49	598
630HC	Row-Crop (Hi-Crop)	1958–1960	Gas, LP, or all-fuel engine.	-	-
730 General Purpose	Row-Crop	1958–1961	First production 08/04/1958. Last U.S. production 06/15/1960. Gas, LP, all-fuel, or diesel engine.	53	593 (LP), 594 (diesel), 605 (gas), 606 (all fuels)
730 Standard	Row-Crop	1958–1961	730 export production continued until 1971 in Argentina. Gas, LP, all-fuel, or diesel engine.	-	-
730 Hi-Crop	Row-Crop	1958–1961	Gas, LP, all-fuel, or diesel engine.	-	-
830	Row-Crop (standard tread)	1958–1961	Also built in Mexico. Diesel (gas or electric start).	73	632 (see 820)
435	Row-Crop	1959–1960	Two-cylinder two-stroke GM engine. First production 03/31/1959; last 02/29/1960.	-	-

1959 MODEL 435

The 435 has a chassis similar to the 430W, but is powered by a General Motors 2-53 two-cylinder, two-stroke, 106-cubic-inch diesel engine. The engine was equipped with an exhaust-gas-driven blower. For you motorheads, bear in mind this is *not* a turbocharger but an exhaust gas-driven impeller used to improve exhaust gas scavenging. The tech is a bit confusing, but the impeller whine gives the engine a distinctively powerful howl. That same GM engine was used in the industrial version of the tractor, the Model 440. Another very similar model was the 445, which was built in Rosario, Argentina. *Mecum Auctions*

CHAPTER FOUR

THE REVOLUTION

NEW GENERATION MODELS

*"Innovation distinguishes between
a leader and a follower."*
—Steve Jobs

MODEL 4010

This battered machine is a survivor and is also the only Model
4010 factory-equipped with a gasoline engine (and, yes, now
retro-fitted with a diesel mill). *Keller Collection / Lee Klancher*

The 1950s farmer was hungry for more power. Hydraulic implements and accessories of the era functioned at higher ground speeds and required more power than past iterations. Pretty much all of the large tractor manufacturers lagged behind this demand, with their largest machines putting out 60 to 70 horsepower in a market where the increasing number of farmers working big acreage needed machines putting out 100 horsepower or more.

In the early 1950s, Deere management determined that new four- and six-cylinder engines were the best way to produce the power the market demanded. Protracted debates were held in boardrooms and on production floors. Not only was this a shift in a core marketing philosophy for the company, the development costs for retooling the entire line were astronomical.

This decision to abandon the two-cylinder was part of a philosophical shift for Deere. It had been content to be the second largest tractor manufacturer by sales

MODEL 4010 FIRST PRODUCTION

The New Generation tractors raised the bar for tractor technology and operator interface. *Keller Collection / Lee Klancher*

volume for several decades. But there was blood in the water in the mid-1950s, and the Deere team had made moves that positioned them to go in for the kill.

Deere's careful cost management gave it profit margins that were some of the best in the heavy equipment industry. Its brand loyalty was outstanding, dealer network solid, and the reliability of its equipment legendary.

The competition, on the other hand, was struggling. While Ford had struck lightning with the Fordson in the 1920s and the N series tractors in 1939, both had been eclipsed by the competition. Ford's 1950s efforts were relatively staid machines, and issues with transmissions had diminished their market standing considerably.

The industry leader, International Harvester, was dragging its feet. It had a major model launch in 1939, and refined that platform well into the mid-1950s. Distracted with ill-fated efforts to move into construction equipment and home appliances and saddled with poor leadership and arrogance, International failed to invest enough into its tractor line. The massive company was also struggling with excessive debt and thin profit margins—problems that would haunt it into the 1980s.

Not long after the new Model 50 and 60 tractors were introduced, Deere decided to create an all-new line of tractors. In April 1953, key John Deere leaders gathered with industrial designers Henry Dreyfuss and William Purcell to set objectives for the new tractor design. The group decided the machine would be new from the ground up and would be more compact than previous tractors, with an uncluttered underside and streamlined rear end.

The engineering staff asked how much they could depart from history for the new design. Charles Wiman responded that the only absolute rule for the new design was green and yellow paint.

With this uncharacteristically open edict from Deere, the design team attacked their drawing boards and went to work creating a new machine.

NEW GENERATION INTRODUCTION

When the new-for-1961 Deere models hit, they were a sensation around the country. This is Machinery Hill at the Minnesota State Fair.

Duane Lundquist / Minnesota Historical Society

As machines became part of everyday society, the science of creating them progressed from improving function to improving the way they interacted with the humans that ran them.

As anyone who has tried to coach their grandfather to turn on his video in Zoom, had arguments with Alexa, struggled to operate a fax machine, had their "smart" tv ruin their day, or had the failure of a single part stop their entire harvest dead can attest, the art of making machines play nicely with their masters is an ongoing challenge.

The earliest tractors were incredibly difficult to start, turn, drive, tune and . . . well, pretty much anything you did with them was hard.

So as early as the 1930s, machine builders began to focus on how to improve not just the machine, but the interaction with humans.

Scientists began to study body measurements in the 1930s and adapted machines accordingly. Pilots were studied fairly intently, and during World War II, it was discovered that a significant number of them crashed due to confusing control systems. After the war, the deep pockets of the defense industry funded studies by teams of engineers, psychologists, industrial designers, statisticians, and more. Human factor engineering was born, resulting in machines that interacted more effectively with humans.

Henry Dreyfuss's ethics tied neatly into this field. At a time when Brooks Stevens pushed planned obsolescence and Raymond Loewy claimed the most beautiful curve was a rising sales graph, Dreyfuss stated his design goals to be ease of use, safety, efficiency, and comfort.

When the first professional organization for industrial designers was formed in 1944, the distinguished group admitted only those who had designed a wide variety of product lines, and Dreyfuss was admitted by consensus.

In the 1950s, Dreyfuss developed standard measurements for people. He gave ranges for everything from finger length to height and dubbed them Joe and Josephine. These two figures became the standards designers would use to create everything from office chairs to steering columns.

The development of the New Generation tractors offered Henry Dreyfuss an opportunity to practice some of his beliefs. He had advocated that John Deere change its simple seat since he started working with the company in 1937. Deere's organization was so fragmented at that time that the suggestion was ignored. Each department head would have had to sign off on a new seat.

So the seats for John Deere tractors continued to be designed in a less-than-scientific manner. In fact, one executive said that seat designs were based on the man with the biggest ass they could find.

The seat would change dramatically beginning in the 1950s with help from Dr. Janet Travell, who had worked with HDA on the seats for the Lockheed Electra and would later become President John F. Kennedy's physician.

Another innovation pioneered on John Deere tractors was the use of universal symbols. Dreyfuss and his wife, Doris Marks Dreyfuss, developed symbols that could be used on equipment and signs around the globe. Some of this work was done at Deere & Company, which was one of the few manufacturers interested in replacing text with symbols. The rabbit and the turtle on the throttle of most John Deere tractors built since the late 1960s are two examples and are included in Dreyfuss's landmark work, *Symbol Sourcebook*. Published in 1972, the book includes more than twenty thousand universal symbols and a multilingual contents page printed in eighteen languages, including Chinese and Swahili.

While today the symbols and colors used on John Deere tractors appear simple and logical and allow anyone to intuitively understand the controls' basic functions, the effort to create them was enormous.

One of the fundamental beliefs of HDA at this time was, "We bear in mind that the object being worked on is going to be ridden in, sat upon, looked at, talked into,

HDA SKETCHES

These drawings by the Henry Dreyfuss Associates show the progression of the New Generation design. Note the changes in colors, model designation, and front-end treatments. In some, you can spot the shape of the early "V" engines Deere considered, while others appear to have an inline engine. Given the fact that the early sketch had a front end similar to the final, and others show variations as late as April 1960, one wonders if Deere management thought the original front-end design to be too plain. Drawings are brush and wash or watercolor, and crayon on off-white tracing paper. *Matt Flynn © Cooper Hewitt, Smithsonian Design Museum*

JD C14-1
JR 1-30-59

JD C14-1
JR 1-28-59

JD C14-1
JR 2-2-59

JD C14-1
JR 1-30-59

JD C14-1
JR 2-3-59

4-14-60

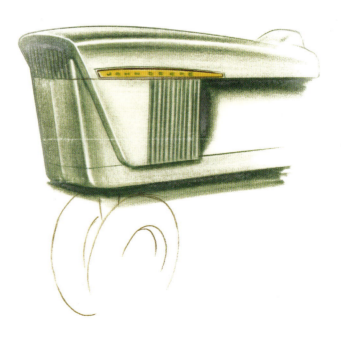

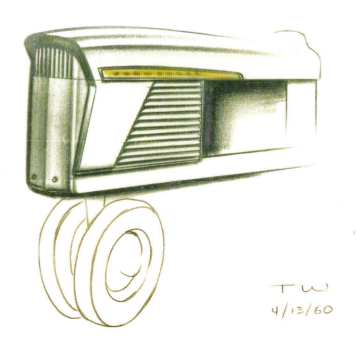

4/13/60

activated, operated, or in some other way used by people. When the point of contact between the product and the people becomes a point of friction, then the industrial designer has failed."

With the New Generation tractors, the agency brought the full force of its understanding to bear. William Hewitt, one of the lead HDA designers on the project, later wrote, "We were completely integrated with the company's engineering and marketing groups . . . the company brought us into action at the very beginning." In a 1961 publication, "Industrial Design: A Vital Ingredient," Purcell spelled out how the HDA designers helped shape the tractor design, and were included in the process as the designs were transformed into working prototypes and manufacturing patterns.

With the New Generation, HDA was a core part of the team. As a result, these machines were designed to fit people in an unprecedented manner.

OX-1 EARLY DESIGN

This is the New Generation hood and front built in wood and clay at the Henry Dreyfuss Associate's location in Pasadena, California. *Chuck Pelly Collection*

NEW GENERATION 10 SERIES

Early in 1953, after years of acrimonious debate, the leaders at John Deere decided to build a tractor engine with more than two cylinders. The two-cylinder had been a mainstay for the line, but everyone from engineering to upper management believed the limits of the technology had been explored and the time had come to change.

The decisions were protected with utmost secrecy, so much so that no records exist of the actual decision to build the new engine (and new tractor).

John Deere was well positioned for the change—it had been the number two tractor builder in the United States for decades and had a firm grip on that position. The conservative company had been content with that for decades, but that was beginning to change in the 1950s.

For the New Generation tractors to succeed, the machines needed to make a quantum leap in technology. John Deere's development cycle is legendarily long to this day, and the New Generation would take roughly seven years from concept to introduction.

MODEL 1010 HI-CROP

This 1010 is a one-of-a-kind machine; a 1010 field-equipped with a John Deere hi-crop kit. *Renner Collection / Lee Klancher*

To be certain this could be done without alerting the competition, the work needed to be done in secret.

Secrets are not easily kept by tractor companies. Manufacturing facilities for competing brands are often located in the same town. The red and green combine plants, for example, were nearly back-to-back in Moline, Illinois, and everyone from managers to line workers would take lunch, smoke breaks, and reconnaissance missions to peer at the nearby plant and see what was shipping out.

Employees also would switch companies, and every locale in a small town, from the schools to the bars to the softball fields, was a place where "red" employees and "green" employees would mingle and swap stories. Like any small town, word about interesting developments had a way of spreading like wildfire.

Agricultural sites where the newest machines were tested were also well-known and frequented by lots of different color manufacturers. During harvest or

planting season at these locations, company spies would watch for prototype machines in the fields. Actual examples of tractor spy skullduggery included crawling underneath the machines in the wee hours of the morning to scrape off metal samples, disassembling parts of machines hidden under canvas covers, and even management pretending to be dealership owners in order to drive the prototypes.

To prevent this type of corporate espionage with the New Generation, John Deere sequestered a small engineering team in the dark corners of a rented Waterloo grocery store on Falls Avenue known to its denizens as the "butcher shop."

The handpicked group of about twenty engineers was led by test and development engineer Merlin Hansen. The engineers would develop the engine in complete secrecy.

In early 1953, Deere & Company rented an old grocery store on Falls Avenue in Waterloo to house

MODEL 1010 STANDARD

The 1010s were designed and built by the Dubuque-based Deere tractor team, and featured a host of improvements, including new gasoline and diesel engines. Utility versions were typically painted yellow, and a crawler model was also offered. *Renner Collection / Lee Klancher*

1965 MODEL 2010 ROW-CROP DIESEL

The 2010 engine was developed by Deere's Dubuque-based engineering team, while the rest of the machine was a collaboration between the Dubuque group and the engineers based at Deere's new Product Engineering Center.
Horstmann Collection / Lee Klancher

the group. "The first thing Deere did was put paper over the windows," engineer Ed Fletcher recalled in *Designing the New Generation*, by Merle L. Miller. Locals suspected a new bar was being built, and they circulated a petition to protest the establishment. Their assumption was fine with Deere & Company management—the purpose of the facility was not to be public knowledge. Donuts were secretly delivered to the back door to avoid attracting attention. A stern general foreman guarded the facility, escorting any interlopers off the property.

Behind locked doors and blacked-out windows, and at times without heat or air conditioning, the

team created the earliest designs for the largest of the all-new John Deere tractors.

The first engine designs were for a V-4 and a V-6, which the sales team desperately wanted. They had been using the fact that their engines were unique as a sales tool since the 1920s, and they wanted to continue to offer that edge.

Unfortunately, the V-engines didn't work, and that track was eventually scrapped in favor of a more traditional inline design.

While the engineers were at work, the CEO of John Deere, Charles Wiman, became ill and passed the reins to William Hewitt in 1954. By 1955, Hewitt

had issued an edict: "Pass International Harvester." He knew that to do so, the company not only needed the engineers to build better products, but it also needed to improve its sales and marketing machine.

The team was given broad parameters developed by upper management and was instructed to create a completely new machine. The multi-cylinder engines were the biggest news, but the hydraulic system, transmission, frame, bodywork, hitch, and control systems were all new as well. Nearly every functioning system on the new tractors was redesigned and improved.

The new tractors' appearance was dictated by function, with the most styling attention given to the hood. The curves were carefully designed so that the tractor looked "right" whether it was working at an angle in a field or sitting perfectly level. This subtlety was felt to be a crucial way to distinguish the machine in the marketplace. The carefully designed curves of the hood were complemented by seamless surfaces—every screw head and panel junction possible was concealed.

Although Wiman had famously suggested that the only carryover design element from the previous machines should be the green-and-yellow paint scheme, on the design board nothing was sacred. Prototype drawings depicted yellow tractors with

1965 MODEL 2010 ROW-CROP DIESEL

Early Model 2010s had troubles with diesel engines starting and the hydraulics systems. The model received a host of upgrades in 1963 and more again in 1965 to address these issues. *Horstmann Collection / Lee Klancher*

1963 MODEL 2010 DIESEL AGRICULTURAL CRAWLER

The 2010 agricultural crawler was equipped with an eight-speed Synchro-Range transmission, and both gasoline and diesel engines were offered. *Keller Collection / Lee Klancher*

inset stainless-steel panels—wild designs for the conservative Deere ethic.

Under that seamless hood would sit the star of the new line: an engine with more than two cylinders.

The program had begun in 1953 with the intention of having finished machines ready to roll out in 1958. As the design of the new machines progressed, it became apparent that redesigning every single system properly would take more time. The introduction of the new line was delayed. In the meantime, the 20 series was "redesigned" as the 30 series.

Finally, the new machines were introduced with tremendous fanfare. More than six thousand people were flown to Dallas, Texas, and at noon on August 30,

1963 MODEL 2010 DIESEL AGRICULTURAL CRAWLER

This machine was formerly owned by Washington State University. While the industrial 2010 crawler was introduced for 1961, the agricultural version of the 2010 crawler was introduced in 1963, and only 159 of the green-painted ag models were built. The Ertl Company based its toy off this example.

Keller Collection / Lee Klancher

1960, the new 3010 tractor was unveiled at the Neiman Marcus store in downtown Dallas. Harold Stanley Marcus, the flamboyant owner of the store, was the master of ceremonies at this private unveiling. Guests found a giant gift-wrapped package near the jewelry counter. Tish Hewitt, the wife of John Deere leader Bill Hewitt, cut open the bow and the package to reveal the 3010, with diamonds taped to its flanks and a diamond coronet on its muffler. Speeches and other introductions to the new machines lasted all weekend. The entire gala, known as Deere Day in Dallas, was a massive party, capped off with a fireworks display at the Cotton Bowl.

Four models were introduced in Dallas: the 3010 and 4010, as well as the 1010 and 2010. John Deere's New Generation may have arrived glittering with style and jewels, but the machine was conceived in an old grocery store in Iowa.

MODEL 1010 & 2010 DATA

Model	Type	Model Years	Notes	HP	Nebraska Test #
1010R, RU, U, W	Row-Crop Utility	1961–1965	Row-Crop Utility, Utility, and Industrial Wheel variants.	36	802 (gas), 803 (diesel)
1010C	Industrial Crawler	1961–1964	9,094 gas and 6,709 diesel built.	36	801 (gas), 798 (diesel)
1010O	Orchard	1962–1965	63 gas and 9 diesel built.	36	-
1010CA	Agricultural Crawler	1963–1964	380 gas and 127 diesel built.	36	801 (gas), 798 (diesel)
1010R, RS	Row-Crop	1963–1965	Row-Crop and Row-Crop Single.	36	-
1010RUS	Special Row-Crop Utility	1962–1965	4938 gas built.	36	-
2010R, H, RU, W	Hi-Crop	1961–1965	Row-Crop, Hi-Crop, Row-Crop Utility, and Industrial Wheel variants.	46	-
2010C	Industrial Crawler	1961–1964	1,136 gas and 8,193 diesel built.	46	829 (gas), 830 (diesel)
2010RUS	Special Row-Crop Utility	1962–1965	5,075 gas built—LP and diesel unknown.	46	-
2010CA	Agricultural Crawler	1963–1965	159 diesel built.	46	-
2010F	Forklift	1963–1965	279 gas and 38 diesel built.	46	-

"It has been a tremendous show extremely well engineered and prepared. . . I'll never see another like it."
— Theo Brown, diary entry about Deere Day in Dallas, August 31, 1960

Major model changes, always costly, can sometimes be devastating because of the time spent on conversion. The Ford Motor Company's overlong shift from the Model T to the Model A was a classic example of this. (All Ford's assembly plants, as well as its overseas factories, were completely closed for more than six months, and the company did not resume full-scale production until more than a year had elapsed; in the process Ford had irrevocably lost a startling amount of its market share.) Deere probably had spent too much time in the development stages of the four-cylinder tractor, but the conversion to its production at the Waterloo plant was spectacularly well done. The Waterloo shutdown lasted only about five months, and the layoffs were staggered in such a way that as one group of employees was laid off another previously furloughed group would come back. By this time the harmonious results of the 1955-56 contract negotiations with the United Automobile Workers had produced an era of close working relations, and employee-relations tensions during the changeover seemed at a minimum.

Downtime for changeover is typically reflected in production time lost on existing models, and the results of 1960 showed this. Sales dropped from 1959's $577.1 million to $511.7 million in 1960, and the net income took a sharp decline from $50.9 million to $20.3 million. Heavy outlays for the machinery and tooling for the new tractors had pushed total company capital expenditures to approximately $40 million in 1960—the net investment of the company, according to the annual report of 1960, jumped from $98 million in 1959 to almost $120 million in 1960. *Forbes* caught the essence of the situation in its article on the new tractors: "As hindsight proved, Deere could hardly have picked a better year than 1960 to make the changeover. . . . The shutdown coincided with a marked falloff in domestic farm equipment sales as farmers felt the pinch of lower income in 1959 and early 1960. . . . Deere did not miss out on a potential bumper year. In fact, its indifferent profit showing was matched last year by competitors who had no such unusual expenses as Deere incurred." In other words, fortunate timing mitigated what might have been a far greater effect. The efficacy of the change still had to be proven by the success in the new models, but it was to this issue that Hewitt and his colleagues now turned with gusto.

"All day long the big planes buzzed in and out of Dallas Love Field. They carried passengers from New York and New Dorp (Pa.), from Paris, France, and Paris, Ill., from Seattle and Sewanee (Tenn.). When darkness fell on Monday, August 29, 1960, more than 6,000 passengers—the biggest industrial airlift in history—had been safely landed." Thus began the *Forbes* article on "Deere Day in Dallas." The next day's events opened with a preview of the entire line, enthusiastically received by the dealers as the tractors stood in a row in a huge open display area in the parking lot of the Cotton Bowl, near the Coliseum, on the outskirts of the central city. Perhaps the most spectacular single event of that day came at the stroke of noon, in downtown Dallas itself. Inside the prestigious department store, Neiman-Marcus, right next to its exclusive jewelry counter, stood a huge, twenty-foot box. Stanley Marcus, the ebullient impresario and president of the firm, walked up to the box and tugged away its wrappings. *Forbes* described the results: "Its contents: a rakish-looking, grass-green, farm tractor. From its sides myriad diamonds (hastily affixed with Scotch tape by willing [Neiman-Marcus] salespeople) twinkled the name of its maker: John Deere." A diamond coronet was even affixed to the tractor's exhaust pipe. Only slightly less spectacular were the fireworks that evening, the Al Hirt band that had been flown in from New Orleans, the Texas barbecue, and other hoopla. Laced through all of this were hard-sell speeches and informal contacts, with Hewitt himself personally acting as master of ceremonies for the evening events and C.R. Carlson, vice

THE LAUNCH

John Deere's New Generation tractors were introduced at Deere Day, a gala event in Dallas, Texas, in August 1960. This Model 4010 New Generation is serial number 1000, the first Model 4010 manufactured. *Keller Collection / Lee Klancher* president of marketing, leading the Deere marketing contingent.

The choice of Dallas itself was noteworthy. Previously the unveiling of new tractor models always had been done in Waterloo (or Dubuque, if built there). They were in the "heart of the Midwest," logical places for the farmer customer; Dallas was not. But it was the *dealer* who was involved in the initial viewings, and dealers could be brought from all over the country to anywhere. Dallas was just a dramatic and offbeat enough choice to whet their appetites. (In more recent years, Deere has made extensive use of dealer "fly ins," shuttling them from all over the country to Moline in Deere-owned and commercial-charter aircraft).

The proof had to be the models themselves, and their enthusiastic acceptance was quick in coming. Over the next crop year their field results were excellent, and though a drought slowed down sales during the last half of 1961, the North American agricultural machinery results were up considerably in that year and the total sales for the company rose to $561 million (with income up to $36 million).

NEW GENERATION

As the Space Age dawned, Deere's new line offered unprecedented features and powers. *Koharski Collection*

John Deere 1010 SERIES
4 cylinders . . . 35 h.p. . . . Diesel and Gasoline

ROW-CROP UTILITY

Selective sliding-gear transmission with 5 forward speeds and reverse; Category 1 3-point hitch; 12.4-28 or 13.6-28, 4-ply rear tires; adjustable front and rear axles. Excellent stability, wide adaptability, 20-inch clearance.

John Deere 3010 SERIES
4 cylinders . . . 55 h.p. . . . Diesel and Gasoline

ROW-CROP WITH DUAL FRONT WHEELS

STANDARD

Fixed tread or adjustable front axle and adjustable rear axle; 16.9-30 or 18.4-30, 6-ply rear tires.

ROW-CROP

Transmission with 8 forward and 3 reverse positions; wide adjustable axle or dual front wheels; adjustable rear axle; power steering; power brakes; Universal 3-point hitch; 13.9-36, 6-ply rear tires. Low-profile Row-Crop Utility also available.

A NEW GENERATION OF POWER

John Deere 2010 SERIES
4 cylinders . . . 45 h.p. . . . Diesel and Gasoline

HI-CROP

Provides 34-1/2-inch clearance; adjustable front and rear axles; 13.6-38, 6-ply rear tires.

ROW-CROP

Transmission with 8 forward positions and 3 reverse; Universal 3-point hitch; 12.4-36 or 13.9-36, 4-ply rear tires; adjustable rear axle; choice of adjustable front axle (above) or dual front wheels. Also available in low-profile Row-Crop Utility model.

ROW-CROP MODEL WITH DUAL FRONT WHEELS

John Deere 4010 SERIES
6 cylinders . . . 80 h.p. . . . Diesel and Gasoline

ROW-CROP

Dual front wheels (pictured) or wide adjustable front axle; Universal 3-point hitch; 15.5-38 or 18.4-30, 6-ply rear tires.

STANDARD

Transmission with 8 forward and 3 reverse positions; fixed tread or adjustable front axle and adjustable rear axle; power steering; power brakes; 18.4-34, 6-ply, or 23.1-26, 6-ply rear tires.

HI-CROP

32-inch clearance; adjustable front and rear axles; 15.5-38 or 18.4-34, 6-ply rear tires.

There were four models in the new line—the 1010, rated at just under 36 PTO horsepower; the 2010, at 46.6 horsepower; the 3010, at 59 horsepower; and the 4010, at 84 horsepower. The three larger models had Synchro-Range transmission, which provided eight forward and three reverse speeds, with synchronizers in the transmission permitting shifting on the go within a range between forward speeds or shuttle-shifting between forward and reverse. The closed-center hydraulic system of the three larger models provided up to three independent "live" hydraulic circuits to serve a rear rockshaft and one or two remote cylinders. The models were equipped with power steering, and the largest two models had hydraulic power brakes. All four tractors had high ratios of horsepower to weight, allowing operation with equipment loads at higher speeds, reduced lugging strain on the engine, and increased efficiency. All four were available with gasoline, diesel, or LP gasoline engines.

Operator comfort had never been a high priority with tractor manufacturers. In the new line Deere pioneered, with the help of a perceptive outside consultant, Dr. Janet Travell (a posture specialist who was later White House physician for Pres. John F. Kennedy), an orthopedically sound "comfort" tractor seat was developed. The enclosed Roll-Gard cab was introduced in the early 1970s.

The two largest models were relatively the most successful, some 45,000 of the 3010 being sold in the four years 1960–63 and more than 40,000 of the 4010 in just the three years 1960-62. Their successors, the 3020 (65 PTO horsepower) and the 4020 (91 PTO horsepower), developed in 1963, were even more spectacularly received. In the period 1963–71, more than 86,000 of the former were sold and in the same period some 177,000 of the 4020 were made and sold. The latter was far and away the most widely sold single model of tractor ever built by the company.

Deere had dubbed the new tractors a "New Generation of Power" and *Forbes* thought this "somewhat

grandiose." Yet, in terms of the models' eventual success and Hewitt's efforts to establish irrevocably a new company image, the pretentious name for the model group and the hoopla surrounding the Day in Dallas seemed to justify the superlatives. *Forbes* called the new models "symbols of a livelier, more dynamic company" and attributed this in part to Hewitt himself. "He has transformed the eminently successful but rather shy and conservative company," they concluded.

1961 MODEL 3010 UTILITY

Produced from 1961 to 1963, the Model 3010 was offered with gas, LP, and diesel engines. Variants produced include Row-Crop, Standard, Row-Crop Utility, Industrial, Special Row-Crop Utility, Lanz Special Standard, Grove, and Orchard variants. Total production was 45,222. *Renner Collection / Lee Klancher*

1962 MODEL 3010 LP

The 3010 is one of the few New Generation LP tractors to have the fuel tank under the hood—a design advocated by Henry Dreyfuss. This is one of 103 3010 LP Standards built. *Mecum Auctions*

MODEL 3010 DATA

Model	Type	Model Years	Notes	HP	Nebraska Test #
3010 11T	Row-Crop	1961–1963	12,525 gas, 2,442 LP, and 23,675 diesel built. First production (gas) 07/12/1960.	-	-
3010 12T	Standard	1961–1963	493 gas, 103 LP, and 3,017 diesel built.	59	762 (diesel), 763 (gas), 764 (LP)
3010 14T	Row-Crop Utility	1961–1963	654 gas, 32 LP, and 1,108 diesel built.	-	-
3010 15T	Industrial	1960–1964	192 gas and 648 diesel built.	-	-
3010 16T	Special Row-Crop Utility	1962	118 diesel, others unknown.	-	-
3010 17T	Lanz Standard	1962–1963	Export to Mannheim, Germany. 139 diesel built.	-	-
3010 18T	Orchard	1962–1963	19 gas and 57 diesel built.	-	-

The dominant concern for farmers in the latter part of the twentieth century was finding a machine with adequate horsepower to help them farm more acres with less help. This need prompted farmers of the 1950s to cobble together hand-welded machines powered by high-horsepower engines salvaged from industrial equipment and farmers of the 1960s to replace 100-horsepower six-cylinders with 300-horsepower V-8 diesel engines.

MODEL 4010

John Deere tractors prior to the New Generation placed the operator behind or above the rear wheels. When the front wheel hit a bump, the shock was transmitted to the operator. The New Generation tractors place the operator in front of the back wheels, improving operator comfort.

Keller Collection / Lee Klancher

The war wages on even today. While the manufacturers of today offer machines with 600 horsepower and more, some use Polish firmware to hack their ECM and get enough power to work massive plots of ground with minimal available help.

In the 1960s, more horsepower meant more sales, and all the manufacturers with skin in the game battled hammer and tong, particularly in the lucrative high-horsepower row-crop tractor category.

Deere won round one with its 4010. While the model featured a progressive design, tremendous features, and solid ease of use (in most systems), the factor that kept the farmers flowing into the showroom floor was a class-leading 72 drawbar horsepower—ten more ponies than the nearest competitor and 37 percent more than the flagship of its direct competitor, the International 560.

Deere's competition scrambled to respond, and all the major tractor makers had substantially revised models out between 1963 and 1965. Many of them, like International's highly capable 806, were competitive with the 4010.

While the competition struggled, John Deere was building more weaponry. It unleashed the revised 4020 in 1963, with 83 horsepower at the drawbar, and took over the number one spot in the agricultural tractor industry the same year.

In the high-stakes arms race of the 1960s, Deere fired often and won nearly every battle that took place.

1960s ROW CROP HORSEPOWER WARS, ROUND ONE

Manufacturer	Year	Model	Engine	Price	(Year)	Nebraska Tractor Test Rating
John Deere	1961	4010	380 CID six-cylinder diesel	$5,500	(1963)	72.58 drawbar hp
Oliver	1960	1800A	283 CID Waukesha six-cylinder diesel	NA		62.55 drawbar hp
Allis-Chalmers	1961	D19	Turbocharged 262 CID six-cylinder diesel	$5,300	(1964)	62.05 drawbar hp
J. I. Case	1960	831C	301 CID four-cylinder diesel	$6,000	(1969)	58.3 drawbar hp
International Harvester	1958	560	282 CID six-cylinder diesel	$5,500	(1963)	53.12 drawbar hp

MODEL 4010 FIRST PRODUCTION

This is the first 4010 built. One early objective with the New Generation tractors was to keep them compact while dramatically increasing horsepower. The 4010 wheelbase is only about six inches longer than the Model 730, and PTO horsepower increased from 56.7 to 84.0. *Keller Collection / Lee Klancher*

MODEL 4010 GAS (ONE OF ONE BUILT)

John Deere made 170 diesel and 11 LP 4010 Hi-Crops. By contrast, they built just one gas-powered unit. This 4010 was that gas model, and it saw years of use on a commercial farm in California. This tractor has not been changed since the Kellers acquired it. Even the yellow markings on the side are original. The number 16 was painted on the tractor at the commercial farm where it was used. According to Deere & Company records, this tractor was originally shipped to Appleton, Wisconsin. Before it was sold, it was sent back to the factory, retrofitted with a diesel engine, and shipped to California. Diesel 4010 serial numbers begin with 22T. This tractor—serial number 23T 36420—retained the 23T of a gas model after the engine was switched. Toy manufacturer Ertl used this very machine to create a scale-model tractor. The resulting model had a gas engine, and Ertl received numerous complaints from knowledgeable collectors, so they changed it. The original scale model is now highly collectible. *Keller Collection / Lee Klancher Keller Collection / Lee Klancher*

MODEL 4010 & 5010 DATA

Model	Type	Model Years	Notes	HP	Nebraska Test #
4010 Row-Crop	Row-Crop	1961–1963	3,613 gas, 4,459 LP, and 36,736 diesel built. First production (gas) 07/08/1960; last 09/24/1965.	84	759 (gas), 760 (LP), 761 (diesel)
4010 Hi-Crop	Hi-Crop	1961–1963	0 gas, 17 LP gas, and 170 diesel built.	-	-
4010 Industrial	Industrial	1961–1963	10 gas, 0 LP, 170 diesel built.	-	-
4010 Special Standard	Special Standard	1961–1963	Export only. 35 diesel built.	-	-
4010 Standard	Standard	1961–1963	98 gas, 792 LP, and 11,370 diesel built.	-	-
5010 32T/T323R	Standard	1963–1965	5,463 Standard, 83 JD700, and 2,007 Industrials built. All diesel. 5010-I renamed JD700 in 1965.	121	828

1963 MODEL 5010 FIRST PRODUCTION

The bruiser of the New Generation was introduced for the 1963 season. This unit is serial number 23T 01000, the first production model built. The Model 5010 uses an eight-speed Synchro-Range transmission. 5010 Standard model years are 1963 to 1965. *Keller Collection / Lee Klancher*

1963 MODEL 5010

The big Model 5010's steering is hydraulic with power assist. The disc brakes are also hydraulically actuated. Its turning radius is just over 12 feet, with the inside brake applied. The 5010 produces 105.92 horsepower at the drawbar. In 1962, field tests at the Nebraska Tractor Test Laboratory found the 531-ci inline six-cylinder diesel produced 121.12 PTO horsepower. *Keller Collection / Lee Klancher*

UNLOADING RICE AUGER CART 1965

This Model 5010 is unloading rice at the Kole family farm. From the book, *The Photo Story of The Kole Farm & Family 1935–1968*

NEW GENERATION 20 SERIES

Every day, Deere leader Bill Hewitt commuted from his home in Rock Island, Iowa, to his office in Moline, Illinois. The ten-minute drive came to a close as he drove his Jaguar Mark X across the glass-enclosed bridge leading to Deere's corporate headquarters.

The complex was designed by famed architect Eero Saarinen, who also designed buildings for IBM, General Motors, and CBS, as well as the Gateway Arch in St. Louis. Completed in June 1964, the pillars on the Deere building were finished in structural steel known as Cor-Ten. This material was shiny for a few years, but anodized over time to a dark, rich brown.

The location is determined by tradition. John Deere—the man—built his first plant in Moline when most of the town's business was run in a single mill. He understood the value of being on the banks of the Mississippi River, and the good local supplies of wood and coal.

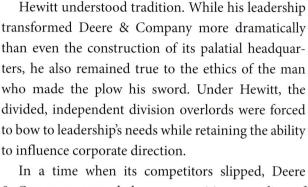

SAARINEN DESIGN

In 1964, Deere opened a new administration building designed by Eero Saarinen. Here Deere leader Bill Hewitt (left) meets with Saarinen to discuss the design in 1956. *Eero Saarinen Collection (MS 593). Manuscripts and Archives, Yale University Library*

Hewitt understood tradition. While his leadership transformed Deere & Company more dramatically than even the construction of its palatial headquarters, he also remained true to the ethics of the man who made the plow his sword. Under Hewitt, the divided, independent division overlords were forced to bow to leadership's needs while retaining the ability to influence corporate direction.

In a time when its competitors slipped, Deere & Company created the most exciting new line of tractors in history. Shortly after that, the company pioneered the development of a rollover protection system, a system that saved thousands of farmers' lives. Deere then decided to offer the patents to any tractor manufacturer, giving the technology it developed away so that it would actually be implemented.

Only two short years after its new line launched in Dallas, Deere's goal of stepping into the number one position came to pass. In 1963, John Deere's annual tractor sales ($762M) exceeded those of the International Harvester Company ($664M) for the first time in history. John Deere tractors represented 34 percent of all the wheeled tractors sold in the United States. Deere's profitability was also the best in the wheeled-tractor industry. By any measure, the company became the world's foremost tractor manufacturer.

Reaching that pinnacle was an impressive accomplishment that resulted from more than a century of growth and development. It is particularly amazing when considering that the company started in the farm tractor industry late and with an established competitor. When IH formed in 1902, the company brought most of the major agricultural companies under one banner, allowing it to dominate the industry in the early part of the century. The company would eventually face a lawsuit for its monopolistic tendencies, but that wouldn't happen for more than fifteen years.

Deere prevailed by attending to its founding principle of producing great equipment. It also stood true to core values of long development cycles and careful cost control. The difference for Deere in the 1960s was a new level of aggression.

VISIONARY MUSCLE

This sketch by HDA staff dated June 5, 1968, blends late 1960s hot rod ethic with the New Generation's dominant impact on the industry. *Henry Dreyfuss Archive, Cooper Hewitt, Smithsonian Design Museum*

The company pressed its advantage mercilessly. The upgraded 3020 and 4020 were introduced for the 1964 model year, only three years after the New Generation set the world on fire in Dallas. The big news on the models was increased horsepower and what amounted to the Holy Grail for 1960s tractor makers: a powershift transmission that could shift the full range of gears under power.

All the makers understood matching speed to load was an essential farm need of the era. However, developing transmission technology was incredibly expensive, and the powershift technology was not quite sorted. Pretty much everyone had some kind of shift-on-the-fly tech, most of it balky and all of it horsepower-consuming. The Ford's was so bad it is now blamed for the demise of the line, and the solutions offered by IH and Case only allowed minor adjustments rather than the ability to shift through the entire range under power.

Deere's Power Shift system worked quite well and was offered as an option alongside the tried-and-true Synchro-Range transmission. The trade-off was you lost about 5 percent of your horsepower to the new Power Shift transmission, which meant this massive investment in new technology was not selling as much as they would like. Deere adjusted for this with some upgrades for the 1969 season that included engine improvements that meant power output ratings were identical for Synchro-Range and Power Shift models.

As a point of comparison, IH would not develop its own powershift transmission until it was merged with J. I. Case in the 1980s.

Quite simply, Deere's investment in farm technology in the 1960s and beyond vaulted it into the number one position and landed it in a position so dominant the competition would spend decades scrambling to catch up.

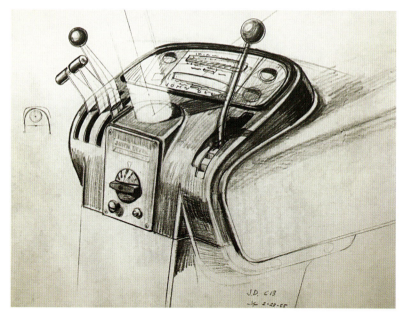

1955 DASH CONCEPT

Jim Conner of HDA drew this in 1955.

Matt Flynn © Cooper Hewitt, Smithsonian Design Museum

1962 DASH CONCEPT

Heavily marked concept for the 20 series and later dashes. Color pencil and graphite on cream tracing paper. *Matt Flynn © Cooper Hewitt, Smithsonian Design Museum*

1961 4010 DASH

This is the dash on the first production 4010.

Keller Collection / Lee Klancher

1964 3020 DASH

This shows the newer style dash, with a Power Shift transmission and LP gas controls. *Keller Collection / Lee Klancher*

1969 4520 DASH

This is a very similar layout, with a manual (Synchro-Range) transmission. *Keller Collection / Lee Klancher*

1967 MODEL 3020 DIESEL

Introduced in 1964, the 3020 pressed Deere's market advantage with more horsepower and an unrivaled list of standard equipment that included power steering, power brakes, and industry-leading operator comfort. *Renner Collection / Lee Klancher*

MODEL 3020 DATA

Model	Type	Model Years	Notes	HP	Nebraska Test #
3020	Row-Crop	1964–1968	Standard, Row-Crop, Row-Crop Utility, Hi-Crop, Special Row-Crop Utility, Orchard, Lanz Standard. Only 11 Hi-Crop gas built. 68,652 total production.	70	848 (DP), 851 (GP), 852 (LP-P), 940 (DS), 941 (GS), 942 (LP-S)
3020 (upgraded)	Row-Crop	1969–1972	Only Row-Crop and Hi-Crop variants. Hydraulic control moved from front to side console. Total production 19,341.	72	1010 (GP), 1011 (GS)

LONG-LIVED HARDWARE

Five decades and change later, many 20 series tractors are still hard at work. Tom Jarvis drives one of those at Springdale Farm, which has been owned and operated by the Jarvis family since 1949. Photo taken on June 29, 2015, near Cherry Hill, New Jersey. *Highsmith, Carol M / Library of Congress*

1964 3020 ORCHARD, FIRST PRODUCTION

Built on September 26, 1963, this is the first 3020 orchard model built.
Keller Collection / Lee Klancher

1964 MODEL 3020 LP HI-CROP

Protruding LP tanks were one of Henry Dreyfuss's pet peeves. He advocated a smooth hood design for the first New Generation tractors. With the Model 3020 LP, however, Deere engineers could not find a way to conceal a tank large enough to fuel a day's work. *Keller Collection / Lee Klancher*

1967 MODEL 3020 HI-CROP

This 3020 was painted yellow at the factory and worked garlic fields at Foremost Gentry Farms in Gilroy, California. The engines in 1969 and later Model 3020s were upgraded with more displacement and power. *Keller Collection / Lee Klancher*

1965 MODEL 3020 LANZ STANDARD

This model was built in Deere's Waterloo plant, equipped with modifications for the European market, and shipped to Mannheim, Germany. This is one of 117 such machines built. The unique plates and gauges of the machine are shown on the right-hand page. *Keller Collection / Lee Klancher*

**1964 MODEL 4020
SECOND PRODUCTION**

Arguably the most popular
John Deere tractor in
history, the 4020 offered
class-leading horsepower,
modern features, a Power
Shift transmission, and
reasonable pricing. This
4020 was the second 4020
built, and the first Power
Shift model. *Purinton
Collection / Lee Klancher*

1960s ROW-CROP HORSEPOWER WARS, ROUND TWO

The fierce fight to be the top dog in the row-crop tractor horsepower wars continued in the mid-1960s. International fought back hard against the 4010, introducing its 806 with a surprising 94.9 PTO horsepower (tested). The John Deere held an edge with its Power Shift transmissions, but it had wisely been working hard on the next update. It introduced the 4020 in 1964 with nearly the same rating and countered the red machine's salvo by upgrading the 4020's horsepower in 1966. Deere upped the ante further with the turbocharged 4320, 4520, and 4620 in 1968.

Deere in the 1960s (and beyond) was relentless in its willingness to pour research and development into burying the competition.

1960s ROW-CROP HORSEPOWER WARS, ROUND TWO

Manufacturer	Year Introduced	Model	Engine	Price (Year)	Nebraska Tractor Test Rating
John Deere	1969	4520	404 CID six-cylinder turbocharged diesel (Power Shift)	$11,600 (1969)	122.36 PTO hp
John Deere	1969	4020	404 CID six-cylinder diesel (Power Shift)	$9760 (1969)	95.83 PTO hp
International Harvester	1963	806	361 CID six-cylinder diesel (TA)	$6,800 (1967)	94.93 PTO hp
John Deere	1964	4020	340 CID six-cylinder diesel (Synchro)	$6,900 (1964)	91.17 PTO hp

1966 MODEL 4020 HI-CROP

Deere's optional Power Shift transmission could move through the entire eight-speed forward or four-speed reverse range while under load. Coupled with refinements to fix minor issues and upgrade others, the new 4020 was the standout model in the segment. *Lough Collection / Lee Klancher*

1967 MODEL 4020

The 4020 was also a critical machine in farm safety, as the roll-over protection system (ROPS) was developed on that model. Deere released the patents in 1966, and Deere's design became an industry standard for open platform tractors. It saved the lives of thousands of farmers. *Renner Collection / Lee Klancher*

MODEL 4020 DATA

Model	Type	Model Years	Notes	HP	Nebraska Test #
4020	Row-Crop	1964–1968	Early style. Power Shift and Synchro-Range Transmission options. Row-Crop, Standard, Hi-Crop, Special Standard, JD600 variants. 129,906 built. Gas, LP, diesel.	95	849 (DS), 850 (GS), 853 (LP-P), 930 (S), 934 (LP-S), 939 (GS)
4020 (upgraded)	Row-Crop	1969–1972	Late style. Improved engine block, pistons, rings, liners; new oval muffler; 12v replaced 24v electrical system; revised controls. Row-Crop, Hi-Crop, and Industrial variants. 55,851 built.	95	1012 (GP), 1013 (GS), 1024 (GP)
4020HC	Hi-Crop	1963–1972	0 gas, 19 LP, and 682 diesel Hi-Crops built.	94	-
4020S	Standard	1963–1972	Received same 1969 upgrades as 4020 Row-Crop. 36 Special Standards with diesel engines, steel seat, short rear fenders built for export in 1962.	96	-

1971 MODEL 4020 HFWA

Deere used several different cab suppliers for the 4020 cabs. The earliest were made by Crenlo, and later models used one supplied by Stolper. In 1971 and beyond, the 4020s were fitted with a Hinson cab, as is shown here with heat and air conditioning. This is also fitted with hydraulic front-wheel assist, which gets poor reviews for performance but high marks for looks. *Harrell Collection / Lee Klancher*

Deere's industrial line dates back to the Model D, and over the years thousands of John Deere tractors were built and sold in a variety of colors to work in factories, the military, and other non-farm enterprises.

Early John Deere industrial models (AI, BI, and DI) were sold with limited success. This makes them rare and highly sought after by collectors today, but it didn't earn much favor with management back then.

Deere continued to sell industrial tractors throughout the 1930s and 1940s. The industrial version of the Model L sold well, as the small machine was well-suited to work on factory floors.

1967 MODEL 500

The Model 500 is the industrial version of the Model 3020. The redesigned cowl and hood give it a powerful stance. This 1967 Model 500 bears serial number 100867 and was painted orange at the factory. *Keller Collection / Lee Klancher*

As demand grew, Deere & Company created a separate division in 1957 to build industrial equipment and launched its new product line that same year in Chicago. The Model 64 all-hydraulic bulldozer came out in 1958. In 1959, the industrial line accounted for $48.2 million in gross sales. As the line increased, sales did the same. By 1969, industrial sales at Deere totaled more than $217 million.

The first New Generation versions of the industrial tractors were simply dubbed 1010 industrial, 2010 industrial, and so on. Later models, however, were given a unique nomenclature. The industrial version of the 1020 was known as the 300, the 2020 was the 400, the 3020 was the 500, and the 4020 was the 600. The 5020 was also sold as an industrial model, dubbed the 700.

1970 MODEL 600

John Deere industrial models were produced since the Model DI in 1925. Deere gave the models unique names with the 440 and made the models more specialized in the mid-1960s. This Model 600 is the industrial version of the 4020, and this one was used by the U.S. Navy. *Keller Collection / Lee Klancher*

1970 MODEL 600

The closed-center hydraulic system developed for the New Generation tractors provides power to multiple systems. It also uses a higher-pressure system, which means that older hydraulic equipment does not function well with it. The braking system on New Generation tractors is an enclosed oil-bathed disc system. These hydraulic brakes represented large improvements in braking power and longevity, though the hydraulic system in the New Generation tractors required an entirely new type of oil, developed by the Lubrizol Corporation. *Keller Collection / Lee Klancher*

JOHN DEERE INDUSTRIAL MODEL DATA CHART

Model	Type	Model Years
DI	Industrial	1925–1941
AI	Industrial	1936–1941
BI	Industrial	1936–1941
LI	Industrial	1941–1946
MI	Industrial	1949–1952
420	Crawler Loader	1957–1958
420I Special Utility	Industrial	1957–1958
440	Industrial	1958–1960
440C	Industrial	1958–1960
830I	Industrial	1958–1961
840	Industrial	1959–1962
1010 C	Industrial Crawler	1960–1964
1010 CL	Crawler Loader	1960–1964
1010 Wheel	Industrial	1960–1965
500	Industrial	1964–1965
600	Industrial	1964–1970
700	Industrial	1965
440	Skidder	1965–1967
350	Industrial Crawler	1965–1970
450	Industrial Crawler	1965–1970
760	Industrial	1965–1970
350C Loader	Crawler Loader	1965–1970
350 Loader	Crawler Loader	1965–1970
450 Loader	Industrial Crawler	1965–1970
400	Industrial	1965–1971
300	Industrial	1965–1973
500A	Industrial	1966–1969
700A	Industrial	1966–1972
440A	Skidder	1967–1970
500B	Backhoe Loader	1969–1971
760A	Industrial	1969–1975
450B	Industrial Crawler	1970–1972

Model	Type	Model Years
350B	Industrial Crawler	1970–1974
301	Industrial	1971–1973
401	Industrial	1971–1973
310	Backhoe Loader	1971–1975
500C	Backhoe Loader	1971–1982
410	Backhoe Loader	1971–1983
510	Backhoe Loader	1971–1983
401B	Industrial	1973–1974
302	Industrial	1973–1981
300B	Industrial	1973–1981
301A	Industrial	1973–1981
450C	Industrial Crawler	1973–1982
302A	Industrial	1974–1981
555	Crawler Loader	1974–1982
401C	Industrial	1974–1983
350C	Industrial Crawler	1974–1985
550	Bulldozer	1975–1982
310A	Backhoe Loader	1975–1982
750	Bulldozer	1976–1984
850	Bulldozer	1978–1984
310B	Backhoe Loader	1982–1986
550A	Bulldozer	1983–1984
450D	Bulldozer	1983–1985
455D	Crawler Loader	1983–1985
401D	Industrial	1983–1986
410B	Backhoe Loader	1983–1986
510B	Backhoe Loader	1983–1986
710B	Backhoe Loader	1983–1988
415B	Backhoe Loader	1983–1993
610B	Backhoe Loader	1984–1986
450E	Bulldozer	1985–1987
450E Long-Track	Bulldozer	1985–1987
210C	Industrial	1986–
610C	Backhoe Loader	1986–

Model	Type	Model Years
350D	Industrial	1986–1988
355D	Industrial Crawler	1986–1988
310C	Backhoe Loader	1986–1991
315C	Backhoe Loader	1986–1991
410C	Backhoe Loader	1986–1991
510C	Backhoe Loader	1986–1991
710C	Backhoe Loader	1988–1993
300D	Backhoe Loader	1991–
510D	Backhoe Loader	1991–
310D	Backhoe Loader	1991–1997
315D	Backhoe Loader	1991–1997
410D	Backhoe Loader	1991–1997
710D	Backhoe Loader	1993–2003
310E	Backhoe Loader	1997–2001
315SE	Backhoe Loader	1997–2001
410E	Backhoe Loader	1997–2001
210LE	Industrial	1997–2007
110TLB	Backhoe Loader	2000–2012
310G	Backhoe Loader	2001–2007
310SG	Backhoe Loader	2001–2007
315SG	Backhoe Loader	2001–2007
410G	Backhoe Loader	2001–2007
710G	Backhoe Loader	2003–2007
310J	Backhoe Loader	2007–2011
315SJ	Backhoe Loader	2007–2011
410J	Backhoe Loader	2007–2011
710J	Backhoe Loader	2007–2011
210LJ	Industrial	2008–2011
310K	Backhoe Loader	2012–2019

WHEATLAND HORSEPOWER WARS

The big wheatland tractor war got interesting in 1965, with the 1206 making a run, but it was outgunned by both John Deere and the surprisingly strong Allis-Chalmers D21 Series II. In the end, the 5020's 141 horsepower made it the king of the two-wheel-drive world in the mid-1960s.

MODEL 5020

The 141-hp Model 5020 was produced from 1967 to 1972. This one is using a Waldon dozer blade and Howard Rotovator in 1966. From the book, *The Photo Story of The Kole Farm & Family 1935–1968*

Manufacturer	Year Introduced	Model	Engine	Price (Year)	Nebraska Tractor Test Rating
John Deere	1969	5020	531 CID six-cylinder diesel	$15,000 (1972)	141.34 PTO hp
John Deere	1966	5020	531 CID six-cylinder diesel	$13,395 (1970)	133.25 PTO hp
Allis-Chalmers	1965	D21 Series II	426 CID six-cylinder turbocharged diesel	$5,700 (1964)	127.70 PTO hp
John Deere	1963	5010	531 CID six-cylinder diesel	$11,000 (1965)	121.12 PTO hp
International Harvester	1965	1206	361 CID six-cylinder turbocharged diesel	$9,450 (1967)	112.64 PTO hp
Oliver	1964	1950	212 CID four-cylinder two-cycle turbocharged diesel	$12,000 (1974)	105.79 PTO hp
Minneapolis-Moline	1965	G708	506 CID six-cylinder diesel	$9,000 (1965)	101 PTO hp (claimed)

XR60 EXPERIMENTAL

With the horsepower wars heating up in the 1960s, Deere experimented with this overpowered two-wheel-drive XR60. According to John Deere test engineer Richard Michael, prototypes were putting out well over 180 horsepower. These machines never saw production. *Richard Michael Collection*

Jon Kinzenbaw learned how to weld just by giving it a try. You could say, in fact, that Kinzenbaw's entire career was founded on his innate ability to transform an idea into a beautifully executed mechanical result. Given his success as a leader in farm technology innovation, he has taken that principle to its very highest level.

As a kid, he applied his welding skills to build himself a mini-bike and then a go-kart. His neighbors had them, but when his father would not buy them for him, Jon built them.

His early efforts were built with wood. His first go-kart was powered by a Maytag washing machine engine with bicycle wheels.

He found the wooden frame and bicycle wheels inadequate for the demands of a young motorhead. The twelve-year-old Kinzenbaw taught himself to weld and built a steel-framed kart with a three-speed Chevy transmission.

His father then refused to buy him a car in high school. Kinzenbaw scraped together his limited funds to purchase two salvaged Fords—a 1956 Crown Victoria and a 1954 Crestliner with a glass top. He mated the good parts from each and drove the car for a year.

His first job was at an International Harvester dealership in Marengo, Iowa, overhauling transmissions and engines. He later worked at a gas station in Ladora, Iowa. With the Vietnam conflict heating up, Kinzenbaw enlisted in the Army Reserves. After basic training, he came back home and decided to put his mechanical skills to work.

With all of twenty-five dollars to his name, he borrowed a little money to buy an old block building in Ladora and open a welding and repair shop. Kinze Welding opened on December 10, 1965.

Not long after opening, while working for a dairy farmer, Kinzenbaw noticed that a four-wheel-drive loader would be helpful to clean out the barn.

"He was a neighbor who was a good friend of everybody's and he had me come and milk his cows," Kinzenbaw said. "He had a 40-head cow herd, Holsteins, and I milked those cows. I observed his problem of not being able to get in the barnyard to get the manure hauled away in the spring because his old tractor was always

stuck . . . I observed his problem and so a year after I had milked those cows for him, I went into business and one of the first projects was to build him a four-wheel-drive loader."

After cooking up a concept, Jon fired up his welder and transformed two Allis-Chalmers WC tractors into a four-wheel-drive loader.

Kinzenbaw doesn't recall making a single drawing. He just conceived the idea with some friends and neighbors, and then built the machine. He was twenty-one years old.

"One thing led to another and when you build something from the ground up, you have to cross all of those bridges as you get to them. . . . I knew enough about differentials and rear ends and how that all worked. I could see and visualize and put that together kind of in my mind. No drawings, nothing like that," Kinzenbaw said.

"If you think about it, one of the tractors' rear ends goes in its normal forward direction, the other one is when it's coupled up has to go backwards.

**CUSTOM LOADER
NUMBER ONE**

One of Kinzenbaw's first projects was to build this loader using two Model WC rear ends and a Chevy engine and transmission to create this custom four-wheel-drive loader in 1966. He built it without any drawings—it was done entirely ad hoc. *Kinzenbaw Collection*

Jon built this incredibly
well-finished custom loader
in his shop with some help
from the customer, farmer
Bill Feller. The machine
was powered by an IH six-
cylinder from an old military
truck. Jon recalls charging
$1,000 to $1,500 for it.
"It seemed like big bucks
in those days," he said.
Kinzenbaw Collection

"Well, then, to add to that, the one that went forward had an engine sitting on it backwards, so it was wrong, and then the other one was wrong because it was going the wrong way. I had to flip the differentials I think in both of those so that they were coupled together and then I used a six-cylinder Chevy engine and a truck transmission and a hydraulic cylinder to steer it and just took it one piece at a time."

That unit came out so well, Kinzenbaw took on more projects like it. He built a custom IH loader, repowered an IH 450 with a John Deere 4020 engine, and constructed three-wheeler "floater" tractors. These early efforts are incredibly well-finished machines.

The quality was not a coincidence.

"One of my things when I was in business, I'd seen farmers that had a welder that should not have had a welder and they'd break something and they'd cobble it together and they'd throw a piece of scrap iron on it and patch it up. Now, it would work but it just looked terrible.

My goal whenever I fixed something . . . I would make the patch look like it belonged on there. In fact, I'd put one on the other side if necessary to make it look right."

"You do a job, do it right and it comes back many times over by word of mouth," he wrote in his book, *Fifty Years of Disruptive Innovation.*

In spring 1967, Kinzenbaw drove up to see a custom-built four-wheel wagon, and stopped in to see Clarence Carlson, who knew of a repowered John Deere 5020 with a 300-horsepower V-8 engine. Carlson told Kinzenbaw of several other farmers who wanted similar projects and explained that those who tried this found the big mill tore apart the stock Deere clutch.

This led to Jon installing a 318-horsepower Detroit Diesel engine into a 5020 for Dave Bystricky of Reinbeck, Iowa. Kinzenbaw solved the clutch problem by installing a double 14-inch truck clutch—which proved to solve only a portion of the problem.

The first was powering the live power take-off, which needed to spin when the engine was running, even if the clutch was disengaged. Kinzenbaw's solution was effective and simple.

"It just kind of fell together and when it was all said and done, it was pretty unique."

The other problem encountered by previous builders was balance, as the other builders used both a John Deere flywheel and the flywheel from the Detroit Diesel engine.

"I just threw all of that away and started over and mine was forked in the center for the most part and I didn't have a balance issue," he said.

The stock Deere transmission, while not rated for 318 horsepower, was able to handle the power just fine. Kinzenbaw believes that Deere transmissions were historically overbuilt because of the power pulses of the old two-cylinder.

"The old John Deere tractor was a tough old cookie, but if you think about it, it's like hitting that transmission with a sledgehammer. Every time that engine crankshaft went around, you got one power pulse.

"When they went to the six-cylinder engine, now they're hitting it three times per revolution, which made it some smoother, but they still overbuilt it. When I put the V-8 Detroit in, it was a two-cycle, which meant it was eight cylinders and each cylinder fired every revolution because they didn't have to have that exhaust and intake stroke. Every time it came up it fired, so we're hitting the crankshaft eight times per revolution instead of one with the old two-cylinder."

Which wasn't to say the 5020 transmission required no changes to handle more than double the horsepower. Fifth gear was used more heavily by the repowered 5020 owners, and was held in place on the stock transmission with nothing but a snap ring, which could fail under heavy load.

"I just took the welder, took the cover off the transmission, reached in and put three quarter-inch stitch welds on top of that snap ring and that eliminated that problem. That's about the only thing we ever did to the transmission or rear end in a John Deere [5020] tractor."

The result was the stock 141-horsepower Deere six-cylinder was replaced with a Detroit Diesel V-8 that could be tuned to up to 318 horsepower (Kinzenbaw noted that some were tuned more mildly, in the 250- to 275-horsepower range). For farmers with a lot of acreage, this meant less time in the fields. As word of Kinzenbaw's machine spread throughout the upper Midwest and into Canada, demand grew. He repowered a variety of New Generation models using the Detroit Diesel engines.

A natural entrepreneur, Kinzenbaw borrowed a little money and expanded his operation, purchasing a new building in 1969. More than two hundred repowered 5020s were built.

To complement the high-powered tractors, Kinzenbaw created larger implements.

At the 1971 Farm Progress Show, a repowered 5020 pulled an adjustable-width seven-bottom plow that Kinzenbaw had custom-built. The adjustability and sheer size of the plow made it interesting. Kinzenbaw sold

rights to it to DMI, White Farm Equipment, and International Harvester. The technology that made the plow adjustable was patented.

The high-powered setup was a hit, and the Kinzenbaw reputation grew, along with it a line of implements and wagons.

When the John Deere 6030 appeared in 1972, the demand for repowers faded a bit.

"The 6030 had a turbocharger and an aftercooler so they could squeeze more horsepower out of it," Kinzenbaw said. "I think the 6030 could quite easily do 200 horse, but it still wasn't enough for the guy that wanted 300 horse."

Kinzenbaw recalled his company repowered very few 6030s, and then moved on.

"We only did one or two as I recall, and we just kind of got out of it. We were into other things, and we had designed the planter and were showing the world how to fold a planter. I just kind of walked away from the repower about 1975," he said.

In the years to come, Kinzenbaw would continue to push the agricultural machine industry with innovative machinery. This hands-on inventor's work repowering John Deere tractors was not quite done.

5020 REPOWER

The project that put Jon's shop on the map was equipping a 5020 with a Detroit Diesel engine capable of 318 horsepower. The demand for these quickly accelerated and put Jon on the path to building Kinze, an innovative agricultural implement and machine building company. *Kinzenbaw Collection*

MODEL 820, 1020, & 2020

In the early 1960s, Deere invested billions into purchasing and establishing plants around the globe. To maximize the advantage of these locations, the company needed common tooling and parts. Harold Brock was tapped to head the engineering team.

A former Ford sales manager, Brock had extensive experience with the Ford N series tractors, a platform that was similar in size and configuration to the tractor Deere needed to sell abroad and at home. When the time came to create a small tractor that could compete in the international market, Brock was a logical choice.

The basic parameters for the new machine were laid out in September 1960. It needed to be a general-purpose tractor with a low profile, with an assortment of models ranging from roughly 20 to 50 PTO horsepower. The machine needed to be easily

1969 MODEL 1020 VINEYARD

Introduced in 1965 and produced until 1973, the Model 1020 was a new design created to meet the needs of small farms worldwide. The 1020 was manufactured both in Dubuque and Mannheim.
Keller Collection / Lee Klancher

configured to different markets and use standardized components.

The company used two platforms as starting points. One was the existing 1010 and 2010 made in Dubuque; the other was the small Lanz tractor built in Mannheim, Germany.

Brock assembled a team of Deere engineers from Dubuque, Mannheim, and France. "Each had their own idea of what a tractor should be," Brock said. "The Germans and the French wanted different speed options. Most people [in Europe] live in the city and go out to the farm in the country."

In the past, drawings had been converted from metric to standard or vice versa. Conversions were invariably rounded off, resulting in parts that didn't fit. This team developed a blueprint that used both metric and U.S. measurements, and employed symbols rather than words. The system became an industry standard.

By 1965, the tractors were ready for production. Sorting out where the machines would be built was almost as difficult as designing the tractors. Tariffs, shipping, and customs had to be considered for each country that would build parts. In the end, manufacture of the machines took place primarily in Dubuque and Mannheim. Other components were built at plants around the world.

Three different models were sold in Europe: the 32-horsepower 310, the 40-horsepower 510, and the 50-horsepower 710. The worldwide tractors gave Deere a better product for Europe, and tremendously

1969 MODEL 1020 VINEYARD

This 1969 1020 vineyard model is narrower than a standard 1020. It also has a hard nose and lights mounted beside the engine instead of on the fenders. This example uses a gas three-cylinder engine good for 34.5 PTO horsepower. *Keller Collection / Lee Klancher*

valuable experience navigating world markets. However, the machines *didn't* give Deere a large market share in Europe.

In the United States, four new models emerged as the domestic versions of the worldwide tractor: the 820, 1020, 1520, and 2020. At home, Deere & Company was settling into its role as a market leader. The rest of the world would have to wait.

MODEL 820, 1020, 1520, & 2020 DATA

Model	Type	Model Years	Notes	HP	Nebraska Test #
820	Utility	1968–1973		31	-
1020	Utility	1965–1973	LU (Low-Crop Utility), HU (Hi-Crop Utility), RU (Row-Crop Utility). Gas or diesel engine.	38	935 (gas), 937 (diesel)
10200U/VU	Orchard/Vineyard	1965–1973	Very few built, exact numbers unknown.	38	-
1520	Utility	1968–1973	LU, HU, O (Orchard).	51	991 (diesel), 1004 (gas)
2020	Row-Crop	1965–1971	LU, HU, RU. Gas and diesel engines.	54	936 (gas), 938 (diesel)
20200	Orchard/Vineyard	1965–1971		54	-

Extraordinary effort transformed the Model 2020 into an orchard tractor. The same can be said of Walter Keller's acquisitions of the rare example shown here.

The 2020 orchard is a bit of a Frankentractor, the creation of an engineering team working midnight magic to get the operator down low beneath the orchard branches. The entire operator's platform was relocated behind the transmission and differential housing; the seat, steering mechanism, and all control levers had to be stretched and reconfigured.

The result is vaguely industrial in appearance—and a bear to operate. The levers are not well marked and understanding which gear the machine is in can be tricky. In addition, the clutch is hand-operated. As any early two-cylinder pilot will report, driving a hand-clutched tractor is no walk in the park.

MODEL 2020 ORCHARD

The 2020 orchard model with a gas engine is a rare machine—the current best estimate is twenty-one were built. This is serial number 84497, which makes it a 1969 model. To make an orchard tractor, the rear controls had to be relocated, which required extensive engineering. *Keller Collection / Lee Klancher*

I started work at the John Deere Waterloo engineering office in May 1960, where a small group of about one hundred staff people were located at the factory. I worked on the 8010 four-wheel-drive tractor. Only one hundred of these were built from mostly purchased components.

That fall, I left to attend Basic Officer Training for the Army Corps of Engineers. When I returned in April 1961, I worked on some production problems on the engines of the new tractors. That winter I was transferred to the John Deere Product Engineering Center.

As a design engineer, I worked on a 4 × 6 foot drafting table and transferred my design ideas by making scaled paper drawings, so some artistic skill was useful. Our basic tools were a drafting square, drafting templates, compass, and a slide-rule. Telephones were shared by four or more people. We had a Bendix computer that was about the size of a refrigerator that had been programed to make gear calculations, bearing life design, and shaft deflection calculations. We would put the input information on data sheets, take it to the computer group, then pick up the results sometime later. Before a computer was available for gear calculations, gear designs were made using Friden calculators. In the late 1950s Deere & Company in Moline had a computer that could calculate gear design.

I was assigned to a new product transmission design group that was working on a project to create the new worldwide tractor models 1020 and 2020. Mike Mack was my design group supervisor. Marion Cessna was working with him on the project. Vern Rugen was leading the product development group. There were two other engineers—George Pakala and Pete Wetrich—and five designers—Bob Reynolds, Roger Jacklin, Karl Freasman,

SMALL 20 TRANSMISSION

The 820, 1020, and 2020 shared an eight-speed forward and four-speed reverse transmission. *Keller Collection / Lee Klancher*

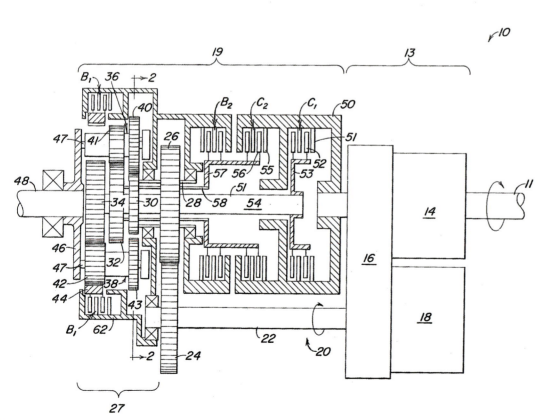

U.S. Patent   Feb. 7, 1984   Sheet 1 of 3   4,429,593

Gordon Marquart, and Ralph Fencl—working on the project. Bob Reynolds created a schematic layout of the transmission case and Roger Jacklin did the gear design.

The Product Engineering Center was devoted to the design and development of new tractor products. There were two large engineering rooms with several hundred engineers, technicians, and support staff. Most of the first designs of the new tractor line were well under way before I joined the group.

My first project was to design a secondary (parking) brake, which was required for the European market. Our Massey-Ferguson competitor used a simple V-block brake on the axle shaft. It took me three months to work out a complex, self-energized, V-block band brake to meet the specifications. I was never satisfied with the design, but other engineers worked on the problem without finding a better solution. I also created a mechanical park system, which involved a complex linkage that actuated a cam to block a pawl into a gear.

We then started on a second design to reduce cost. I suggested a revised parking brake system that used tapered spline. My supervisors accepted my suggestion, but we had to provide an interlock to ensure one of the range gears was engaged to provide park. This led to my first of eleven patents (3,301,078) that used a pin and five quarter-inch balls to prevent movement into the park position unless one of the range gears was engaged. When Bob Woolf from Dubuque saw my plan he said, "Look at all those little balls!" It was a really simple design, and this was my first patent!

I was then asked to try to reduce the cost of the dual-stage clutch that released the traction clutch on the first stage and then released the PTO clutch on the second stage. I developed a simple stamped lever that had one drilled hole instead of a cast lever that required some machining, two holes, and a third tapped hole. In addition, our original clutch concept tended to overheat, so I put fins on the pressure plate to promote cooling. Also, the pressure plate bolt durability was poor, so I used my favorite durability solution. Rather than change the bolt, I reduced stress concentration on the bolt.

Later in the development of the transmission, the first design of a hi-lo planetary transmission flew apart when one of the clutches failed, so it was decided that we had to create a different design. By this time, I had become one of the group's key firemen and worked with Gordon Marquart to develop the two-piston Power Shift hi-lo design and shared my second patent (3,352,166) with Gordon.

THE HYDRO-MECHANICAL TRANSMISSION DEERE NEVER BUILT

By Richard Michael

In 1970, International Harvester introduced the Farmall 1026 hydro with a dual-range hydrostatic transmission. After it was introduced, I helped look at a more efficient limited range hydro-mechanical concept to be competitive.

Since a full hydrostatic transmission would be very inefficient, we chose to use a small-sized variable hydraulic pump and constant displacement hydraulic pump to provide variable speed split with a direct mechanical path and followed by a four-speed forward and one-speed reverse transmission. This was my fourth patent (3,736,813) with Jim Kress and Lyle Madson. A prototype of the limited-range hydro-mechanical transmission was built but was not produced because the Farmall 1026 was too inefficient to be successful as a farm tractor and was discontinued in 1972.

1968 MODEL 2510 GAS HI-CROP

Released for the 1966 season, the 2510 mated the new 53-horespower engine designed for the upcoming 2020 to the 3020 chassis. While this model has a Synchro-Range transmission, the 2510 hi-crop gas was also built with an optional Power Shift transmission. This example is one of only seven 2510 gas high crops, and one of only four that were equipped with the Synchro transmission. *Keller Collection / Lee Klancher*

1968 MODEL 2510 GAS HI-CROP

The 2510 was a bit of a mash-up, mating a 2020 engine to a 3020 chassis. *Keller Collection / Lee Klancher*

Model	Type	Model Years	Notes	HP	Nebraska Test #
MODEL 2510 & 2520 DATA					
2510 711P (Power Shift)	Row-Crop	1965–1968	15,114 (all variants) 2510s built.	50	913
2510 711R (Synchro)	Row-Crop	1965–1968		54	914
2510 713P (Power Shift)	Row-Crop	1965–1968		51	915
2510 713R (Synchro)	Row-Crop	1965–1968		55	916
2510 731P (Power Shift)	Hi-Crop	1965–1968		-	-
2510 731R (Synchro)	Hi-Crop	1965–1968		-	-
2510 733P (Power Shift)	Hi-Crop	1965–1968		-	-
2510 733R (Synchro)	Hi-Crop	1965–1968		-	-
2520	Row-Crop	1968–1972	Hi-Crop available. Synchro-Range and Power Shift. 6,318 of all variants built.	61	992 (DS), 993 (DP), 1002 (GP), 1003 (GS)

MODEL 2520 HI-CROP

When Deere made
improvements to the 20
series line in 1969, it applied
much of them to the 2510
and changed the name to
the 2520. The engine on the
Power Shift models received
a displacement and power
bump, which kept the power
ratings equal for Power Shift
versus manual shift models.
Purinton Collection /
Lee Klancher

MODEL 4000

Deere lightened the 4020 chassis, applied cost-savings measures to other systems, and used a simplified Synchro-Range transmission to create the Model 4000, which offered class-leading horsepower at a price competitive with offerings from the (ahem) tractors of another color. *Purinton Collection / Lee Klancher*

MODEL 4000 DATA

Model	Type	Model Years	Notes	HP	Nebraska Test #
4000	Row-Crop	1969–1972	Diesel Synchro and Power Shift, gas Synchro and Power Shift (1970–1972). Only 9 gas Row-Crop with Power Shift built. 8,095 built (all variants).	96	1023 (diesel)
4000 Low-Profile	Row-Crop	1972	21 diesel with Synchro-Range and 25 diesel with Power Shift built. No gas.	96	-

1972 4000 LOW-PROFILE SYNCHRO

Released for one year only, this model packed all the features of the 4000 into a lower and more compact machine. A 3020 front axle and steering column were used. This rare example is one of twenty-one made in this configuration. *Smith Collection / Lee Klancher*

MODEL 4320, 4520, & 4620

By Jim Allen

In 1961, Allis-Chalmers was the first company to produce a turbo diesel tractor, the D19. Though the result was appreciated and admired, the other tractor manufacturers took their time developing their own turbo diesel tractors. When needing more power, the manufacturers, and perhaps the American farmer, still had a preference for big-displacement, slow-turning, naturally aspirated diesels. Consider, too, that the old diesel designs still in use in the 1960s did not lend themselves to turbocharging. That wasn't true of John Deere.

MODEL 4520 FWA

The John Deere 4520 was built in 1969 and 1970. This 1970 example is equipped with rare front-wheel assist and a Power Shift transmission. The 4520 was listed at $11,600. Roughly 7,800 units were sold. *Keller Collection / Lee Klancher*

John Deere used 381- and 531-cubic-inch engines in the 4010 and 5010. When the 4010 became the 4020 in 1963, the 381-cubic-inch six-cylinder engine was given a displacement boost to 404 cubic inches. The stage was then set for a new tractor.

John Deere was a company that generally did not allow itself to be rushed into putting new developments on the market before it was ready. In fact, John Deere had benefited when other companies had done so. Although some may have clucked about John Deere being a little behind the others in offering a turbo diesel tractor, the folks at John Deere were confident they had not stepped on their crank when they finally did.

The 4520 debuted for the 1969 model year as the company's first turbocharged diesel tractor. It fit between the very popular naturally aspirated 4020 and the much larger 5020, with its 531-cubic-inch six-cylinder engine. Some have called the 4520 a "4020 on steroids," but that's not precisely true. It used a 404-cubic-inch engine that was based on the same NA engine used in the 4020, but the power train and the tractor were larger and heavier. A more accurate description might be to call it a downrated 5020 rather than an uprated 4020. Unlike some of the other turbocharged tractors on the market, the 4520 wasn't just a tractor to which a turbo was added. The entire

MODEL 4320

Introduced in 1971 to offer a higher-horsepower row-crop tractor, the 4320's power boost came from the addition of a factory turbocharger. While the machine is based on a 4020 chassis, improvements throughout complement the engine's boost to 115 horsepower. This rare bird is one of three 4320s built by Deere with yellow paint. *Smith Collection / Lee Klancher*

package had been built around the extra power, and John Deere used the ad line "Turbo-Built."

By 1969, the 404 six-cylinder had become one of John Deere's mainstays. It was, and is, a highly effective power plant that many think was seminal in cementing the company's success in the 1960s and 1970s. The turbo 404 boosted the PTO power output to a modest 122. The compression ratio was lower than the NA version (15.7:1 vs 16.5) with stronger pistons that used keystone rings and had uprated piston oil cooling jets. The turbo engine had a beefier new block with more main bearing support and improved oil flow.

The most whiz-bang part of the new 4520 was the air filtration system. Designed to both clean the inlet air in a superior fashion *and* offer a long maintenance interval, the air cleaner had twelve venturis that removed backed up dirt. Along with this, a special muffler was designed to siphon off roughly 90 percent of the incoming dirt and blow it out the stack. Pretty clever, huh? Unfortunately, this turned out to be a place where John Deere might have tripped over its crank. It didn't take long for problems to start—dirt would blow into the engine, and in some cases even set the air filters on fire. John Deere had to pay for upgrades to customers' tractors even after the normal warranty had expired. Whoops! With an upgrade, the 4520 became a tractor to admire.

The 4520 was available with either the standard Synchro-Range, or with the Power Shift that offered

MODEL 4820 CUSTOM

This one-off custom creation mates a 1972 4620 with a 40 series cab, 466 engine, hydraulic front wheel assist, and more. The meticulous work by owner Steve Moster creates a gorgeous machine that allows you to wonder what might have been.
Moster Collection / Lee Klancher

clutchless shift on the fly. A heated and air-conditioned cab was available as well as a fiberglass canopy that attached to the ROPS. Later in the run, an FWA system was offered as an option. Some of the last 4520s built in 1972 have been seen with the Sound-Gard cab that was one of the major contributors of the hoopla to John Deere's "Generation II" program.

The 4520 lasted until 1970 when it was replaced by the 4620, which was very similar but aftercooled and made 135 horsepower.

That same year, the 4320 debuted, and it was indeed a "super" 4020. The model used the 404 cubic-inch mill with improved fuel injection and engine

cooling stuffed into a beefed up 4020 chassis. The only transmission offering was a Synchro-Range.

This line would evolve into the even more powerful 4630 of 1973, which made 150 horsepower with an intercooled 404 turbo engine. The 1973 model year was John Deere's second new beginning—akin to the model's 1960 rebirth—and the company debuted a great deal of new technology. Many say that the 1973 Generation II era marks the beginning of John Deere's decades-long domination of the agriculture market. The 4520 and its siblings were major stepping-stones into that era.

MODEL 4320, 4520, 4620, & 5020 DATA

Model	Type	Model Years	Notes	HP	Nebraska Test #
4320	Row-Crop	1971–1972	All equipped with diesel engine and Synchro-Range transmission. 21,485 built.	115	1050 (DS)
4520	Row-Crop	1969–1970	6,213 Synchro-Range and 1,662 Power Shift built.	120	1014 (DP), 1015 (DS)
4620	Row-Crop	1971–1972	5,045 Synchro-Range and 1,883 Power Shift built.	135	1064 (DP), 1073 (DS)
5020	Standard	1967–1972	7,103 Row-Crop and 5,806 Standard built. Industrial version Model 700.	141	947, 1025

CHAPTER FIVE

THE FOUR-WHEEL DRIVE EMERGES

THE 8010, WAGNER, & 7X20 MODELS

"Form follows function."
—Henry Dreyfuss

WA-14

The Wagner models are one of the most interesting (and brief) chapters in Deere tractor history.

Renner Collection / Lee Klancher

239

In 1950, rainy weather hammered Norman County in northwestern Minnesota. Oswald Dallenbach was the Minnesota Extension Agent for nearby Clay County, and more than a decade later, he still recalled how a four-wheel-drive tractor made news in the area. In *Beyond the Furrow*, historian Hiram Dirache recounts Dallenbach's tale of how the farmer's Massey Harris Challenger four-wheel-drive tractor was able to work his land before any of his neighbors could venture into the muddy fields.

Four-wheel-drive tractors were not a new concept. Dozens of different concepts for them had been developed in the 1920s, the wild days of tractor experimentation. Deere's Dain Tractor was conceived and tested during this time and was one of the more refined designs of the era.

By the 1950s, in parts of the country where farms were big, soil was thick, and hired help scarce, the demand for a high-horsepower, four-wheel-drive was significant. The big manufacturers were focused on building machines for the sweet spot in the market—the 40- to 70-horsepower row-crop tractors.

In fact, the 1940s were the golden era of row-crop machines, and the big builders were raking in cash at an unprecedented rate selling 50-horsepower row-crop machines to the average farmer. It's fairly understandable that they ignored the pleas for more power and traction coming from a handful of big-acreage

farmers in the Red River Valley in the Midwest, the hill-plowers in eastern Washington and Oregon, or other areas with vast expanses of rich soil.

Farmers being farmers, when the big manufacturers failed to solve their problems, they took matters into their own hands and welded together a variety of Frankenstein creations.

The stretch to envision and build a heavy four-wheel-drive tractor wasn't as far as you might imagine. As Steiger's first sales manager, Earl Christianson, astutely pointed out, 200-horsepower machines were moving dirt to build highways. Observant farmers quickly figured out that, with a few modifications, the big machines they saw building highways and moving dirt could work farmland.

In Western Minnesota, Alan Adams built a four-wheel-drive tractor using a 401 GMC V-6 for power and two Allis-Chalmers WD-9 rear ends for the drives. His son, Paul, said the tractor was at work in 1956. Multiple examples of similar machines live in private collectors' hands.

Innovation like this was taking place all around the United States, most of them home-built rigs that served their masters with varying degrees of success. A few extremely talented builders, however, turned their Frankenstein creations into gold.

The most notable of these was Elmer Wagner and his brothers in Portland, Oregon. The boys were surrounded by farm country peppered with big farms and heavy soil—conditions that demanded more power. They developed a four-wheel-drive tractor, which was patented in 1953, and formed the Wagner Tractor Company. In 1954, Wagner offered three four-wheel-drive models for sale, all powered by diesel engines with 114 to 165 horsepower.

To put the power in perspective, Deere's most powerful machine in 1954 was the new Model 80, which produced a class-leading 65 belt horsepower. Several of the competitors were close to that, but all paled in comparison to the Wagner.

The Wagners sold well, and the machines played key roles in firing the imaginations of several

WAGNER TR-14

The Wagner tractor played a key role in driving industry innovation with four-wheel-drive machines in the 1950s. *Peter Simpson Collection*

dreamers and manufacturers. The most successful dreamers were the Steiger family. Douglas Steiger and his two sons had a good-sized farm in northwestern Minnesota and ran industrial equipment on the side to make ends meet. They saw a Wagner tractor in 1956, and immediately recognized the machine was a game-changer. If they could replace their two small tractors with one big one, they could free up a man to work and bring in badly needed cash.

Douglas went to his local banker and said he needed to borrow just over $20,000 to buy a Wagner. He added that for about $10,000, he thought he could build a machine. The banker gave him the smaller amount and told him to get busy welding.

The family did just that, and a brand was born in the Steiger family barn. By the mid-1970s, Steiger was one of the leading four-wheel-drive tractor manufacturers.

Farmers weren't the only ones to take notice of the Wagner machines. In 1956, Deere did a test comparison of a Wagner TR-9 and J.I. Case LA, putting the machines to work with a 24-foot chisel plow. Deere president William Hewitt attended the demonstration, indicating the weight the company placed on this area. They began a development program, looking to build a machine to fit into this interesting market.

As was typical of International Harvester in that era, it showed up to the party late and reacted by belatedly tossing money on the fire. In the process, however, it compiled some interesting information about the era.

In 1959, word of Wagner sales made it to the top of the IH heap, and executive Brooks McCormick personally visited the Portland Harvester office to find out how sales were going for Wagner in the region. While there, the local agents collected information on Wagner sales.

While sales near Portland amounted to a handful of machines, IH employee H. W. Berry was able to coax a Wagner employee to cough up intel, and learned Wagner's 1959 production was between 300 and 400 units and was expected to be 600 to 800 units in 1960.

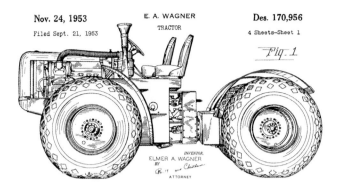

Nov. 24, 1953
Filed Sept. 21, 1953

E. A. WAGNER
TRACTOR

Des. 170,956
4 Sheets—Sheet 1

Fig. 1

INVENTOR.
ELMER A. WAGNER
BY
ATTORNEY

PIONEERING PATENT

Elmer Wagner filed this patent in September 1953.

U.S. Patent

Another IH 1959 report surveyed farmers, county extension agents, and specialists in farm management and agricultural engineering. The results showed that a significant portion of American farmers were interested in more horsepower and four-wheel-drive tractors. Of farmers who owned 70–80 horsepower tractors—the largest of the time—35 percent of them were interested in a four-wheel-drive machine.

Representatives spoke to farmers and discovered that the Wagners were replacing crawlers for two main reasons: one, the machines sat up higher than crawlers, so there was less dust; and two, farmers often had fields that were as much as twenty miles apart, and the wheeled Wagners could be driven on the road and the crawlers would have to be trailered.

With all this good information in hand, IH prompted to go ahead with the ill-conceived 4300, a machine built by its subsidiary, Hough, that had a lot of horsepower (300) and really nothing else to offer working farmers. The machine made no sense to anyone as it was too big, and the company quietly buried the multi-million-dollar project with only forty-four examples built and sold.

R. I. Throckmorton, a product planner assigned to the four-wheel-drive market project, also reported there were rumors of a John Deere 100-horsepower, four-wheel-drive tractor expected to be on the market for 1960.

John Deere's early efforts in the four-wheel-drive market were started sooner and were much better thought-out. Despite this, its path to creating a successful machine for this market had plenty of bumps in the road.

MODEL 8010 & 8020

At a 1959 John Deere field day in Marshalltown, Iowa, a brand-new machine shocked the crowd: a 200-horsepower, four-wheel-drive behemoth called the John Deere 8010.

The powerful machine had features uncharacteristic for Deere.

For starters, the engine was not a John Deere two-cylinder, but a 425-cubic-inch six-cylinder two-stroke Detroit Diesel. The transmission was a Spicer nine-speed, with Clark axles and Westinghouse air brakes. The tractor also featured a shocking $33,000 price tag.

The outsourced bits and sudden appearance made it clear this tractor was a specification build, which meant an engineer drew up the design using as many off-the-shelf parts as possible. This allowed for a new model to be created rapidly, but required compromises based on the available parts and pieces.

John Deere favored long development cycles—think of the fourteen years it took to make the Model D as a classic example. However, the off-the-shelf parts on the 8010 indicates the machine was created with a short development time. The most likely presumption is the machine was a response to the Wagner tractor—and given the fact Deere tested one in 1956, that theory holds a lot of water—or at the worst the increasing market demand for high-horsepower machines.

The engineering team assigned to build the largest addition to the New Generation line started their task by asking farmers what they wanted in a large tractor. In those surveys, farmers responded that they wanted to plow, till, and harrow quickly (at speeds between 4 and 4 1/4 miles per hour) with the largest implements available (which required 5,200–9,500 pounds of drawbar pull). Labor was scarce, so the machine had to be operable by one man.

Farmers were working large plots of land by this time, many of them leased and separated by miles of farm roads. The tractor had to be road legal, meaning no more than 96 inches wide with a tread weight limited to 18,000 pounds. High transport speeds were also important to traveling farmers.

Long hours were the norm for these farmers, so a fuel supply that allowed twelve hours in the field was key. Powerful broad-beam lighting systems would help farmers work around the clock when necessary.

The engineering team transformed specifications into blueprints and drawings into preproduction test mules. The path chosen was the easiest way to quickly build a new machine, that is, sourcing existing components.

The machine that emerged would become the largest, most powerful John Deere tractor ever built: the 1960 Model 8010. Four-wheel-drive supplied the required drawbar power and speed. Maximum drawbar power was generated with 67 percent of the weight distributed on the front wheels. Articulation made the tractor maneuverable in fields and on roads.

Lead engineer Franklin C. Walters spoke about the process of building the 8010 to the Society of Automotive Engineers in 1960. He recounted his words when he and his team presented their machine to the executives at John Deere: "Gentlemen . . . the Model 8010 tractor will go out into your field and work day

MODEL 8010

The 8010 weighs roughly 20,000 pounds and is powered by a General Motors six-cylinder two-cycle diesel engine that puts out 150 drawbar horsepower. *Wisconsin Historical Society / 129043*

in and day out with the same few cents per acre that you've always known with John Deere equipment."

Walters was proud of his machine, but the results did not stand the test of time.

The first issue was the retail price in excess of $30,000, which was wildly more expensive than other machines at the time. This was more expensive than the Wagners by about $8,000. More importantly to people walking into a John Deere dealership, the price was nine times that of a new Model 830 and six times that of a new 4010.

The testing time was clearly inadequate. The Spicer truck transmissions in the 8010 failed if put under anything more than modest loads, a flaw that would have appeared in a normal test cycle.

The articulation joint was also troublesome, and some complained the joint would flex enough to rattle the entire machine and the operator as well.

The transmission was so bad that the few machines that were sold—and Deere only built one hundred 8010s—were recalled and an entirely different transmission was installed, along with a host of other improvements.

The upgraded machines were rebadged the 8020, and Deere took four years to sell them off.

Walter and Bruce Keller bought the 8010 seen here from a dealer in New Hampton, Iowa. They restored the machine and, in the summer of 2009, took it to a show in Illinois. After the show, Gerald Mortensen contacted them. He had heard about the machine at the show, and several details about it sparked his interest. He wanted to come to the Keller farm to see it. When he explained that he was a retired John Deere engineer, the Kellers readily agreed.

At the Kellers', Mortensen walked around the machine a bit. He carefully examined mounts, gauges, and two loose hydraulic fittings poking from the engine cowl. He concluded that the tractor the Kellers purchased was not just any 8010, but serial number OW41A, the prototype

THE ONLY 8010

This Model 8010, serial number OW41A, was a prototype used by the John Deere Engineering Research Division. Former Deere engineer Gerald Mortensen identified a number of distinguishing features on this prototype. The hydraulic couplings protruding from the engine bay were used for a dozer-control system built in March 1961. It was the only dozer control system built. The prototype was converted for industrial use in 1961, and later worked at the Waterloo factory.

Keller Collection / Lee Klancher

that he and his team had built in 1959. The tractor was also the 8010 that appeared at Deere Day in Dallas in 1960. After that event, the prototype was eventually moved on to industrial testing.

The tractor had been heavily field-tested and used for implement shows in Arizona and Iowa, and was mounted with a dozer blade and other modifications that Mortenson used to identify it at the Keller farm. "Deryl Miller and myself loaded this tractor and some other equipment on a flatcar bound for Chandler, Arizona, for a field test with big prototype equipment from other Deere factories in

early November 1959, almost before the paint was dry," the retired engineer said.

Mortenson thought the machine was long gone. "I last saw OW41A with the ATECO low-bowl scraper sitting in a bone pile at the back of the Deere foundry in the mid-1980s," he recalled. "I thought it was headed for the foundry cupola and didn't expect to see it again."

How the machine made it from the scrapyard to the Iowa dealership is a mystery. Mortenson, for one, is just glad this historic piece of iron found a safe home.

8010 DASH
This unique "Bat-O-Meter" voltmeter was installed on OW41A as a trial and is the predecessor of the modern voltmeter instruments now in common usage. Note the factory inventory tag on the dash. *Keller Collection / Lee Klancher*

Only one hundred production Model 8010s were built, and ninety-nine of those were converted to 8020s by Deere. *Fischer Collection / Lee Klancher*

1969 JOHN DEERE WA-14

The WA-14s were built by Wagner-FWD and branded with John Deere paint and logos. The model was powered by a Cummins 12.2-liter six-cylinder diesel rated for 220 horsepower, with a ten-speed transmission and planetary final drives. Only twenty-three of the John Deere WA-14s were built and sold. *Renner Collection / Lee Klancher*

WAGNER

As the 1960s wore on, Deere did not have a successful entry into the high-horsepower, four-wheel-drive market. It was developing its own four-wheel-drive, and the market was heating up. For reasons not entirely clear today, it opted to purchase several models built by Wagner, coat them in John Deere paint and decals, and sell them. Several historians have speculated Deere did so to make sure its dealers had a four-wheel-drive to sell while the new 7020 development was finished. Details of the deal between Wagner and Deere and the motives for the purchase are incredibly sketchy, and the archives are not willing to share, so the story is muddy at best.

Here's what is known.

The story begins in 1922, when Eddie Wagner and his six brothers built a mobile concrete-mixing machine near Portland, Oregon. They called their

creation the Mixermobile, and branched out to build forestry equipment, wheeled shovels, mining equipment, and other four-wheel-drive machines.

One of the boys, Elmer Wagner, designed the first four-wheel-drive articulated vehicle in 1949, supposedly modeling it after one he saw while serving in Europe during World War II. Note that the six Wagner boys went their separate ways, with some building mining equipment, and others involved in timber land grabs.

According to the website, *Unusual Off-Road Locomotion*, Elmer also built an innovative four-wheel-drive, all-terrain, amphibious vehicle—the Go-Devil—in 1961.

Anyway, in 1953 and 1954, three of the Wagner brothers—Walter, Elmer, and Irvin—designed and built a high-horsepower farm tractor, which was introduced in 1954. This company offered four models, the Wagner TR6, TR9, TR14, and TR24, ranging from 110 to 300 horsepower. The machines were

1969 JOHN DEERE WA-17
The other Deere-branded Wagner model was the WA-17, which featured a Cummins 14.0-liter six-cylinder diesel engine rated at 250 horsepower. *Renner Collection / Lee Klancher*

1969 JOHN DEERE WA-17

Only twenty-eight of the John Deere WA-17s were built and sold. Wagner tractors were discontinued entirely when Deere stopped distributing them.

Renner Collection / Lee Klancher

hand-built and constantly refined and improved, so much so that some say that no two models are alike. As was described in the opening to this chapter, these groundbreaking and well-built tractors inspired innovation in the agricultural industry.

In 1961, Wagner sold (or leased) its tractors and logging machines to FWD, a company based in Wisconsin that built four-wheel-drive fire trucks. The tractors were rebranded Wagner Agricultural, the paint color was changed to yellow, and the model names started with WA. The line expanded to six models ranging from 98 to 300 horsepower, the WA-4, WA-6, WA-9, WA-14, WA-17, and WA-24.

In 1968, the Raygo Corporation in Minnesota made a deal to produce at least one (and maybe all) Wagner tractor models. Raygo built construction equipment and had been founded in 1964. The only model to appear with "Raygo Wagner" on it was the big WA-24.

Deere comes into the picture as well in 1968, at which point the historical water gets even murkier. On New Year's Eve 1968, Deere and Wagner signed an agreement that allowed John Deere to sell one hundred WA-14 and WA-17 tractors. Wagner would supply the machines, and Deere would paint them, apply decals, and market and sell them as its own.

While one would expect that the Wagners spent that blustery New Year's Eve toasting a deal likely to ensure a prosperous 1969, signing that document was the most catastrophic moment in Wagner tractor history.

The contract gave Deere the rights to purchase one hundred Wagner tractors, but it ended up buying (and eventually selling) only twenty-three WA-14s and twenty-eight WA-17s.

After dismal sales in 1969, Deere introduced its own four-wheel-drive tractor in 1970 and discontinued the WA-14 and WA-17.

The kicker was that the agreement signed on December 31, 1968, contained a clause that stated that if Deere stopped purchasing and selling Wagner machines, Wagner could not build a competing four-wheel-drive articulating tractor for five years.

That was a straight-up death knell for Wagner. Five years with no sales equals no four-wheel-drive tractor line.

Many questions arise from this odd business deal, and some of the implications are not flattering to Deere. Why was this clause written in? The Deere employees who wrote it surely knew that they had a new line of four-wheel-drives coming shortly, and this clause could spell doom for Wagner tractors.

Why was the deal signed on New Year's Eve? One would assume there was a good reason to have the deal done in 1968.

Why did Wagner sign such a horrible contract? That five-year clause was something any two-bit lawyer or halfway sharp executive would have spotted.

Given that Wagner made a deal with Raygo in 1968, and then another with Deere at the last minute of 1968, my bet would be that even with the tractors selling, cash was not flowing as it should at Wagner. Cash flow is a common problem for heavy equipment manufacturers, and desperation might indeed lead someone to sign a contract to sell one hundred machines at the last second, even if the signing meant the end of the company if the deal fell apart.

I might go so far to speculate that Wagner signed it knowing it was a death warrant, but the ag machinery was doing so badly that it wouldn't survive without the Deere sale.

All speculation, and only the Wagner and Deere folks who signed that contract on New Year's Eve know the correct answer.

And, FYI, the Big Bud and Rite tractor companies were formed as a result of the vacuum created when Wagner tractors were no longer available.

In any case, this perplexing chapter in Deere history created fifty-one rare machines for collectors and fans to obsess about and a fascinating story to ponder.

MODEL 7020 & 7520

After two strikeouts with the 8010/8020 and Wagner WA models, Deere hit one out of the park with the new-for-1971 Model 7020 four-wheel-drive articulated tractor.

First, the decision to design its own machine in-house was the right one, and not an easy one to make. Four-wheel-drive tractors were still a new technology in the mid-1960s, with a small niche market for them.

While the owner surveys done by the major manufacturers clearly showed the demand for more traction and horsepower to allow farmers to more quickly work fields, sales of these machines had not yet become huge.

Development of new tractors is always an expensive proposition, and four-wheel-drive systems and articulation were both new areas to agricultural tractors that required major investment.

Deere was incredibly aggressive with research and development in the 1960s and beyond. It had experimented with component-building the 8010—which was a disaster in terms of the product and the high retail price—and outsourcing to another company, Wagner—which was a disaster for everyone involved, especially the dead and buried Wagner tractor line.

Deere wisely spotted the opportunity in the market for an economical, modestly powered four-wheel-

drive capable of working row crops (meaning it had adjustable wheel widths and adequate crop clearance).

The specifications were a smart choice, and slotted the machine neatly in-between the high-priced, high-horsepower Steiger and Versatiles and the low-priced, budget-friendly J. I. Case machines. With a Deere reputation and pricing closer to J. I. Case than the high-end Steiger, the new Deere four-wheel-drive was a pretty sure bet.

**1971 MODEL 7020
SERIAL NUMBER PLATE**

1973 MODEL 7520

The 531-cubic-inch six-cylinder turbocharged diesel in the 7520 was good for 175.8 PTO horsepower when tested at Nebraska in June 1972.

Renner Collection / Lee Klancher

The engine was sourced from the 4520, their 404-cubic-inch turbocharged and intercooled six-cylinder. Power was boosted a bit in the 7020, and the engine produced 146 PTO horsepower. That was mated to an eight-speed forward, two-speed reverse Synchro-Range transmission with an optional hi-lo range feature.

Development took longer than expected, and several knowledgeable historians have speculated that the delay prompted Deere to panic-buy the Wagners, but for the moment that fact is lost to history.

Two-Cylinder magazine printed an interesting release dated March 25, 1970, that was sent to Deere dealerships, offering a sneak peek at the new 7020. The specifications—minus the price—were included, and the release promised 7020s to begin shipping in August 1970. The first shipment of a 7020 was November 12, 1970, so that March memo was a tad optimistic.

Base price for the new model was $15,975—roughly half of the cost of the overpriced 8010, and not too far off from the pricing of the two-wheel-drive 4620. The price was also well below the cost of comparable machines from Steiger and other four-wheel-drive specialty builders.

Criticized for being complex to service and somewhat fragile, the 7020 was nevertheless a solid seller in the four-wheel-drive class. The upgrade from the 7020 to the 7520 came only a year later, with a larger displacement engine giving the later model a welcomed power boost.

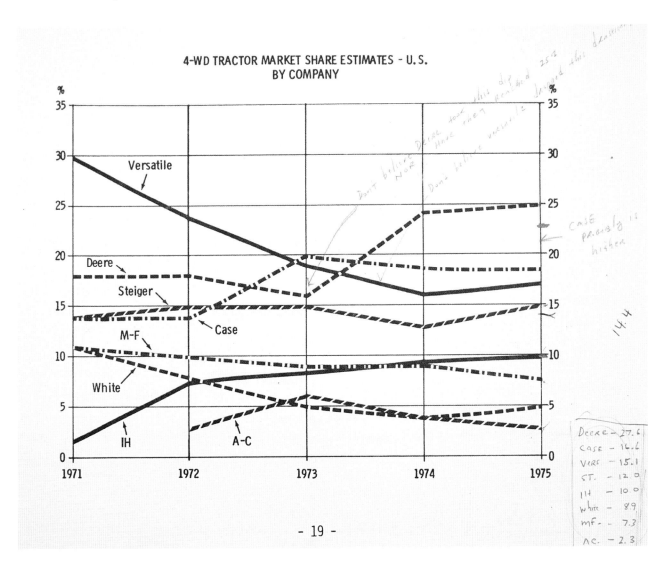

- 19 -

BIG DEMAND FOR TRACTION AND POWER

The four-wheel-drive market grew rapidly in the early 1970s. Sales of four-wheel-drive tractors above 100 horsepower increased from 28,728 units in 1971 to 62,081 units in 1975. This chart shows how John Deere dominated the market with its 7020 and 7520. *Wisconsin Historical Society*

GOLDEN AGE

GENERATION II TRACTORS
1972–1994

"The difference between a successful person and others is not a lack of strength, not a lack of knowledge, but rather in a lack of will."
—Vince Lombardi

1974 6030

Deere roared into the early 1970s with a muscular new line.

Salmon Collection / Lee Klancher

CLAY MODELING GENERATION II

HDA industrial designer Chuck Pelly at work on clay models. The drawings behind him are the four-wheel-drive and two-wheel-drive machines. Pelly said they put flashlight bulbs in the cab to use light to show the ground visible to the operator. *Chuck Pelly Collection*

Deere surprised the agricultural world in the 1960s by stepping up to become the market leader in farm equipment. While its top competitor responded vigorously, the red leader's arrogance and sloppy management had cost it. The sales of the stunning New Generation tractors pushed Deere into the leadership position in the market in 1963.

Harvester's response was a strong counterpunch. Its best efforts of the time were the 706 and 806, which were smartly engineered, high-horsepower, row-crop tractors. In normal times, that would have been enough to level the playing field for a bit and give the former market leader enough leverage to perhaps retake the industry lead.

Deere appeared to have anticipated that, and counterpunched quickly with the 20 series with more horsepower and a Power Shift transmission. The Power Shift (the ability to change speed and power on the fly) was one of those holy grails of agricultural technology, something that was widely in demand and complex and expensive to develop.

At least in the 1960s, the 20 series was the nail in the coffin for Harvester's hopes at regaining the top spot in the industry.

Deere was just getting warmed up.

**GENERATION II ¼ SCALE
WOOD MODEL**

Model of the early
Generation II design in 1969.
*Henry Dreyfuss Archive,
Cooper Hewitt, Smithsonian
Design Museum*

In the twenty years that followed, the company invested intensely in research and development. It not only maintained its leadership position—it distanced itself from the field so much that the impact would be felt for decades.

It's easy to point to the New Generation introduction as the time where John Deere asserted its market domination, but those machines were only one line of revolutionary machines. Many companies have built a new machine, or a line of them, that represented a turning point in farm technology. In fact, most of the tractor manufacturers of the mid-twentieth century created at least one revolutionary machine or line.

In rare cases, a single machine or line could lead to market domination—the most dramatic example of this being Ford with the Fordson and later with the 9N.

Ford was in a unique position, with a brand-new technology desperately in need of innovation that could be used by a very large group of customers. Opportunities like that were long gone by the 1960s—

most farmers had tractors, and the number of farmers in the world had shrunk exponentially.

With Generation II, Deere embarked on a merciless path of creating innovative new machines that continuously set the industry standard. One of the interesting portions of this path is it did this by charging into a dynamic market, as the number of farms worldwide plummeted and tractors needed to be larger and more powerful. The only way for tractor manufacturers to justify the massive research and development expenses was to serve a global market.

This was a difficult period for heavy equipment manufacturers to navigate, and many of the tractor manufacturers would be consolidated or dissolved in this harsh environment.

John Deere thrived.

The period from the early 1970s to the mid-1990s was the time when the John Deere company as we know it today—a goliath that has led its industry for more than six decades—was formed.

MODEL PHOTOGRAPHY

Modeling was key to design in the 1970s, and this is how the photographs of 1/8th scale models were made. *Chuck Pelly Collection*

HIGH AIR-FLOW CONCEPT

This four-wheel-drive had a oval front grille designed to improve airflow. This was not produced. *Chuck Pelly Collection*

CAB EVOLUTION

The cab design evolved from a box to have curved glass. The glass had to be curved to accommodate the Deere's long clutch throw, and the curved glass turned out to be stronger than flat glass and improved visibility for the operator. The round cab concept took this idea to the extreme, but was rejected. *Chuck Pelly Collection*

VISIONS OF THE FUTURE

50-X FUTURE SIX-WHEEL-DRIVE
Designers were encouraged to draw fanciful concepts. Chuck Pelly drew this one in anticipation of the increasing demand for high-horsepower 4WD tractors. *Chuck Pelly Collection*

JOHN DEERE ELECTRIC CAR

The head of the lawn and turf division, a Mr. Hoffman, asked Chuck Pelly, "What about an electric car?" This sketch is the result, and shows Pelly's affinity for compact European design, developed when he went to college in Sweden.
Chuck Pelly Collection

MODEL 830, 1030, & 1530

MODEL 830

The 830 was built in Mannheim, Germany, and powered by a 35-horsepower three-cylinder engine. *Koharski Collection*

MODEL 1030

Also built in Mannheim and powered by a three-cylinder engine, the 1030 was rated for 48 horsepower. *Koharski Collection*

MODEL 1530

The 1520 had been built in Dubuque, Iowa, but the 1530 was built in Mannheim. *Koharski Collection*

MODEL 301A TURF TRACTOR

This industrial model was built from 1973 into the 1980s. This unusual tractor was sold in 1984 by Mid-Atlantic Equipment in Maryland. The tractor was painted with DuPont turquoise from the factory. This one was sold in 1984. *Keller Collection / Lee Klancher*

THE 30 SERIES

By Ryan Roossinck

In the late 1960s and early 1970s, America's farming landscape was growing very rapidly. Farmers were planting more ground, which meant spending more time in the field than ever before. As John Deere watched this expansion unfold, it realized that the needs of the farmer were changing.

Farmers needed more power, a more versatile transmission, and most importantly, more comfort. The power and transmission were the easiest of those needs to meet, but the last one took some unconventional thinking.

The primary objectives for the Generation II tractors were operator comfort and safety. With farmers spending more time on a tractor than ever before, safety was a very real concern. Bouncing around on a loud open station tractor took its toll on the body, and fatigue causes accidents. The new line of tractors needed to isolate the operator from the environment as much as was possible.

To accomplish this, the engineers in Waterloo completely re-imagined the concept of a cab. No longer would they look at a cab as a metal box with windows. For the 30 series tractors, the cab became an integral part of the tractor itself. It needed to shut out the dust, noise, and vibration in order to minimize operator fatigue. The concept was so different that John Deere refused to even use the term "cab" in its advertising! It wasn't a cab—it was a Sound-Gard *body*!

The Sound-Gard body truly redefined the tractor, and it changed farming forever.

NEW MIDRANGE 30 SERIES

In August 1972, Deere announced a new series of tractors featuring more power, new features, and the brand-new Sound-Gard body. *John Deere Archives*

1974 4030

Six-cylinder gas or diesel engines were available in the 4030, with both rated for 80 horsepower. The base model was priced at $17,721 in 1977, the last year of production. *Lee Klancher / Renner Collection*

1973 4230 HI-CROP

The 4230 was available with gas or diesel six-cylinder engines. Both were rated for 100 horsepower. *Renner Collection / Lee Klancher*

1974 4230 LP

The low-profile variant features an underslung exhaust and lower dimensions, making it ideal for orchard or other work requiring lower clearance. *Renner Collection / Lee Klancher*

1974 4430 HI-CROP

The 4430 was powered by Deere's 6.6-liter six-cylinder diesel rated for 125 horsepower. The hydraulic front wheel drive was an available option that offered minimal power to the front wheels and was prone to fluid leaks.

BUILDING THE MOCK-UP

Chuck Pelly is building a mock-up cab on the 4020 above to introduce his new design to Deere executives, circa 1970. The VW in the photograph is the one he would run over. "I had to tie the batteries on with cable, " Pelly said. "When I hit the VW the batteries broke and spun and sprayed acid. Henry Dreyfuss and the Deere executives ran to get behind trees." *Chuck Pelly Collection*

EARLY CAB MOCK-UP

This was the mock-up done prior to the demonstration. This one was mainly cardboard and tape. "It was even flimsier than the one on the demonstrator 4020," Pelly said. *Chuck Pelly Collection*

266   JOHN DEERE EVOLUTION

Chuck Pelly, a well-known industrial designer, penned designs for BMW, Porsche, Samsonite, the Disney monorail, and many more. He went to work in the late 1960s for Henry Dreyfuss Associates (HDA) and was one of the last designers to work closely with the old master himself, Henry Dreyfuss.

One of Pelly's projects while at HDA was the design of the new John Deere Sound-Gard cab. After months of design work, the first mock-up of the cab was to be presented to John Deere executives on a stage. Pelly recalls taking great pride in the new design.

"[The mock-up was] made out of paper, and tape, and bent plastic," he said, recalling it was just a visual representation, and not terribly sturdy. "I'm not a good tractor driver, but I wanted to present it moving."

When the audience gathered, Pelly intended to drive the machine out for them to see it for the very first time.

"I had all the big John Deere people there down at the bottom of the presentation area.

Pelly had very little experience driving a tractor, and the machine got away from him and went careening off track, toward the parked car.

"I ran over the front of the Volkswagen," Pelly said. "Of course, it came apart."

He ran it over with one big right tire. The small car came apart like a cheap watch. Pelly also ruptured the fuel tank.

"I drilled a hole in the tank and the battery spun. Everyone hid behind trees."

Pelly figured that was the end of his work with John Deere and HDA for that matter.

"I packed up my stuff and said, 'Oh, okay I know I'm fired.' Instead, the head of John Deere engineering said, 'We have a job for you. We want you to be a rollover test engineer.'"

Pelly laughed recalling the engineer's good sense of humor. The Sound-Gard cab became a hit for John Deere, and Pelly would continue to work happily with the company and would eventually found his own firm doing work for BMW and many others.

HENRY

Henry Dreyfuss inspects a new fender design. *Chuck Pelly Collection*

SMASHING SKETCH

In this sketch, industrial designer Chuck Pelly recreated the chaos that ensued when he unveiled his new design for the Sound-Gard cab to Deere executives. *Chuck Pelly*

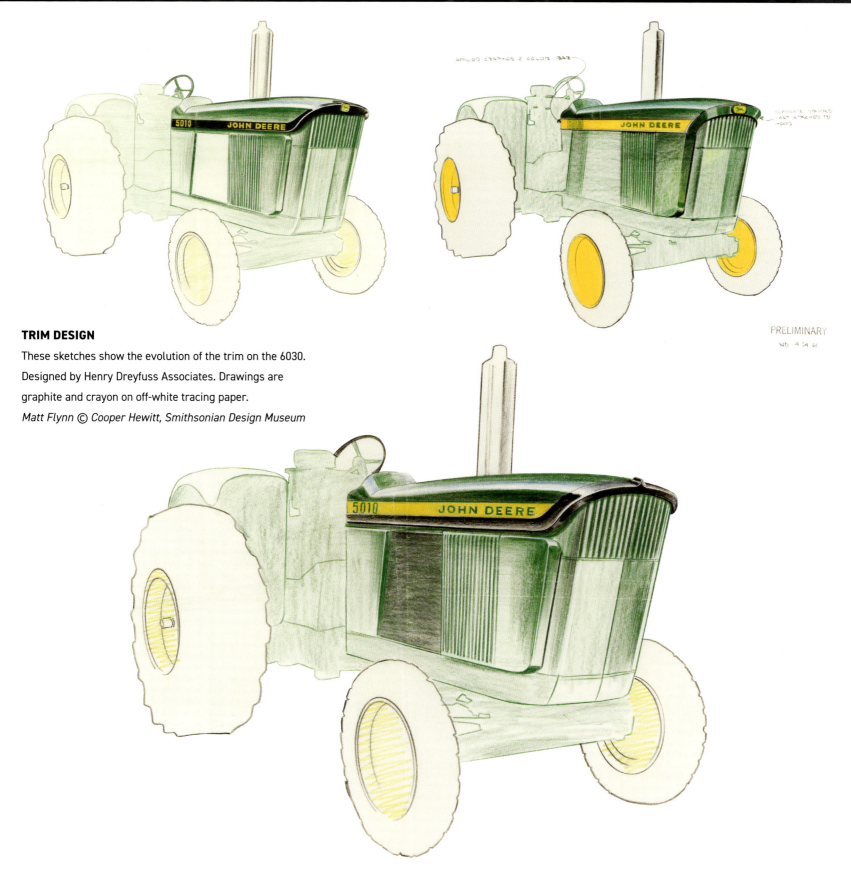

TRIM DESIGN

These sketches show the evolution of the trim on the 6030.
Designed by Henry Dreyfuss Associates. Drawings are
graphite and crayon on off-white tracing paper.

Matt Flynn © Cooper Hewitt, Smithsonian Design Museum

PRELIMINARY

MODEL 6030

1972 6030 FIRST PRODUCTION

With minimal electronics, an eight-
speed Synchro-Range tranny, and
a turbocharged and intercooled
531-cubic-inch diesel good for 175
horsepower, the 6030 caters to power
farmers of the 1970s as well as
millennial farm boys of the modern day.
This 6030 is the first one built, an early
production model used for testing that
has several key differences from the
later production versions.

Walk Collection / Lee Klancher

1973 6030 NON-TURBO

Seventy-five of the 4,028 6030s were equipped
with an optional non-turbocharged 141-horsepower
version of the 531 engine. This is one of them—
note the lack of a housing on the engine.

Walk Collection / Lee Klancher

One of the most significant agricultural technology advances of the twentieth century was the rotary threshing combine. The concept emerged not long after the turn of the century, and a wide variety of dreamers and schemers did their best to turn this idea into a practical machine.

Curtis Baldwin was perhaps the most famous of these, and he would die a broken man after dedicating most of his life to solving the puzzle of how to separate grain from chaff in a rotating cylinder.

While the concept of rotary separation is fairly simple—spin the grain in a slotted tube and let the harvest spit out one way and the undesirable material another—practical application is a multi-variable equation of almost unfathomable scope.

Actually getting a rotating cylinder to effectively separate grain was horrifically difficult to engineer and required millions of hours of refinement and dollars of development.

Consider the fact that any single type of grain—let's take wheat—can vary in moisture content, weight, stalk size, the thickness of the outer layer, and so on. The air itself is critical to rotary threshing, and air varies in humidity and temperature. As these factors in the air change, the rotary separation process changes.

So your rotary thresher has to adapt to all these conditions, and that's just for one crop. You also have corn, soybeans, flax, rice, sorghum . . . each with different weights and properties.

In the first half of the twentieth century, at least a half-dozen well-documented attempts were made to develop a successful rotary thresher by some of the smartest minds and largest companies of the time.

All of them failed.

John Deere experimented with rotary combines. According to an article in *Green Magazine*, Deere's dive into this technology began in 1957 with the XCC-1 experimental, based on a pull-type Model 65. A second experimental, the XCC-5 was built on a self-propelled Model 95 chassis. Deere recorded frustrating problems with plugging and high grain loss into 1960 and 1961. By 1962, the rotary threshing prototype XCC-15 had 125 horsepower, with performance reportedly comparable to the Deere Model 105 combine. Development was dropped not long after that model was built.

Note that International Harvester had a similar rotary program at that time and its engineers experienced very similar frustrations. IH test engineer Dave Gustafson reported that during development in the mid-1960s, his team spent literally eight months locked in a garage analyzing grain flow with a threshing cylinder mounted under glass and a high-speed movie camera recording flow patterns. Gustafson reported they had made zero improvements in the process over the eight months, but had "learned a lot."

New Holland was the first to offer a commercial version of the technology, introducing its TR70 rotary threshing combine in 1975, and IH followed with a rotary combine, the Axial-Flow, in 1976.

The emergence of these prompted John Deere to task one of its top engineers, Dr. Glenn Kahle, to pick up

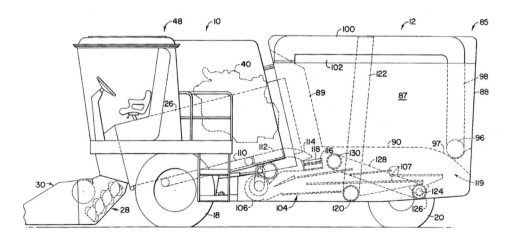

U.S. Patent   Mar. 2, 1982   Sheet 1 of 5   4,317,326

FIG. 1

ARTICULATED COMBINE PATENT

Deere & Co. filed this patent in March 1979 and filed a more detailed articulated combine patent in April 1982. The design was not produced. *U.S. Patent* where they left off in the 1960s and develop a competitive product. Dr. Kahle told the story of his experiences developing that combine in a taped interview on September 2, 2014.

In 1976, Kahle was the manager of the Advanced Harvesting Systems for John Deere, and was stationed in a temporary office in Moline, Illinois. Because of the proximity to the International Harvester plant, Kahle remembers closely monitoring development of its rotary combine.

"We watched the Axial-Flow being built, since IH and Deere were relatively close to each other, like a mile apart," he said. "Whenever they shipped something out, we could see it go by on the railroad car and we'd take pictures of it."

He also said that because the two engineering groups were so physically close, they would interact much more so than other engineers. "There was a lot of interchange because ASA, the American Society of Agriculture engineers, had meetings [in Moline]," Kahle said. "The combine group would intermingle a lot more than the tractor groups would just because they were all in one place and they all lived in the same community. They were much closer than any other part of Deere and IH."

The group Dr. Kahle led, Advanced Harvesting Systems, was a sort of think tank for new technology that

XW-4 EXPERIMENTAL COMBINE

This drawing by HDA is labeled the R10 Harvester and is dated March 1978. *Henry Dreyfuss Archive, Cooper Hewitt, Smithsonian Design Museum*

included John Deere management and engineers from each of the factories in North America and the United Kingdom.

When the International Axial-Flow combine emerged, the John Deere team approached a Minnesota farmer who bought one of the first IH 1460 combines sold. Deere wanted to purchase the machine for testing, and the farmer sold it to them after some negotiation. Rumor has it he was given a brand-new Deere combine along with a handsome sum of cash, but only the local dealer and the farmer know the truth of what went on. In any case, Deere took the freshly minted 1460 to its experimental farm for testing.

Dr. Kahle and his team took the model apart and began work on engineering the next evolution. They developed some fascinating variations on the design. But, as Dr. Kahle recalls, they built their prototype right on top of the IH machine.

"The 1460-like machine actually used a physical 1460," Dr. Kahle said. "We'd bought one, tore it all apart,

ARTICULATED COMBINE MODEL

A concept for Deere's articulated combine created by HDA. *Henry Dreyfuss Archive, Cooper Hewitt, Smithsonian Design Museum* put the connections on it, took the grain tank off, and put sheet metal on it and so forth."

The combine was cobbled together, with the IH 1460 rotor bolted, welded, and screwed to a variety of John Deere off-the-shelf bits and hand-fabricated hardware. Kahle said the Deere engineering team improved on the design considerably, and took it another step beyond with a radical development.

The green-and-red hybrid machine was articulated. The grain tank was removed and put into a cart behind the machine. The modifications resulted in equal weight spread on each set of wheels.

"When [the grain] dumped into the tank, it went into this tube that went back to the grain cart, so you had a grain cart that was low, center to the ground, and would carry twice as much as the tank on the combine. We had four-wheel-drive hydrostatic [to] all wheels," Kahle said. "It was a hell of a machine."

When asked whether this kind of thing was likely, former IH combine engineer Don Watt said such a thing was likely. He also added that his understanding was Deere spent more on rotary combine development in the 1970s than IH did for its entire program—and that Deere was not threatened by the New Holland or the International rotary machines. Dr. Kahle concurred.

"My recollection is that investment [required to build the new combine] was $120 million," Kahle said. "The damn thing was going to have to sell for over $100,000 at that time to make as much money as the 8800, 6600, and 7700," Kahle said. "The profit on those things was just unreal. I don't remember the exact numbers, but our goal always was fifteen percent and my recollection, it was like thirty or thirty-two . . . from a financial standpoint, it was the right thing to do."

Stories of new technology developed by Deere and never built abound. It was founded fiscally conservative, and that ethic has held. It consistently spends massive amounts on R&D—the company also consistently bases its decision whether or not to produce the new technology on fiscal basis.

This is at times frustrating for farmers who would like the new machines, or even engineers who dedicate their lives to creating things that don't make production for decades if ever.

But business is business, and in the latter part of the twentieth century, no one was better at the agricultural equipment business than Deere.

ADVANCED HARVESTING PROGRAM

In the 1970s, another group led by Deere engineer, Dr. Gordon H. Millar, developed a Biomass Converter that converted corn residue to electricity. The projects never made it past the prototype stage, but it was shown to President Jimmy Carter in 1979.

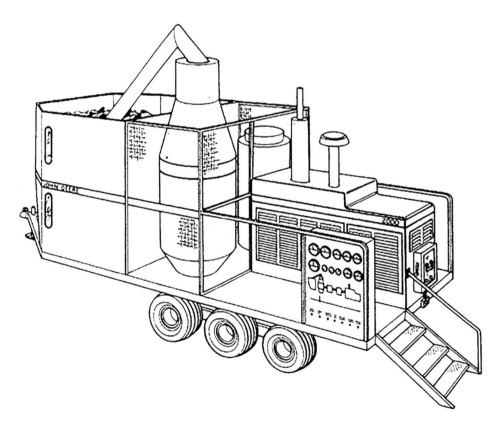

BIOMASS CONVERTER

This sketch is of a working biomass converter built by Gordon Millar, a prolific Deere research team leader who was part of the same advanced engineering group that built the articulated combine. The machine converted corn residue to electricity. *Kahle Collection*

MODEL 8430 FIRST PRODUCTION

John Deere replaced the 7020 and 7520 with its new high-horsepower, four-wheel-drive models, the 8430 and 8630, in 1975. This is the first 8630 built.
Keller Collection / Lee Klancher

1974 MODEL 8430

The 8430 was powered by a 466-cubic-inch turbocharged diesel engine good for 175 horsepower as well as a partial Power Shift transmission, a chassis that used the engine as a stressed member, and a host of new features.
Renner Collection / Lee Klancher

MODEL 8630

The 8630's 619-cubic-inch engine was a turbocharged, intercooled six-cylinder diesel hooked to a partial Power Shift transmission.
Rory Day / Classic Tractor Magazine

30 SERIES DATA

Model	Type	Model Years	Notes	HP	Nebraska Test #
830	Utility	1973–1975	830 also built in U.S.	35	-
830	Utility	1974–1979	Larger engine (2.7L).	35	-
930	Utility	1974–1979	Overseas offering only.	43	-
1030	Utility	1973–1979		48	-
1130	Utility	1974–1979	Overseas offering only.	53	-
1530	Utility	1973–1975		45	-
1630	Utility	1973–1979		59	-
1830	Utility	1973–1979	Same as U.S. Model 2030 and exported to Canada.	60	-
2030	Utility	1971–1975	Built in Mannheim and Dubuque, Iowa.	68	-
2130	Utility	1973–1977		79	-
2630	Utility	1974–1975		70	1157
3130	Utility	1973–1979	Redesigned 1975 front, hood, and decals. Sold as 2840 in U.S.	87	-
4030	Row-Crop	1973–1977		80	1111
4230	Row-Crop	1973–1977	4230 Hi-Crop, 4230 LP (Low-Profile).	100	1112
4430	Row-Crop	1973–1977	4430 Hi-Crop.	125	1110
4630	Row-Crop	1973–1977		150	1113
6030	Row-Crop	1972–1977		175	1100
8430	4WD	1975–1978		178	1179
8630	4WD	1975–1978		225	1180

Jon Kinzenbaw's business repowering tractors had peaked with the 5020 and dropped off when Deere offered the high-horsepower 6030. His business shifted toward building innovative wagons and implements, most for large operations. He also constructed Big Blue, a 20-ton twin-engine, four-wheel-drive tractor that coupled a 6030 with a salvaged 5020.

When the new high-horsepower, four-wheel-drive 8430 and 8630 were introduced for the 1975 season, Kinzenbaw's father-in-law purchased one . . . which put Kinze back in the repower business.

"My father-in-law had a new 8630, and we discovered that he had a bad engine," Kinzenbaw said. Deere offered a replacement engine for about $5,000.

8640 REPOWER

Kinze Manufacturing began repowering the 8430 and 8630 with Cummins diesel engines. The modifications included a new frame under the engine and a tilt-up hood. *Kinzenbaw Collection*

They soon discovered that engine failures were a persistent problem with these models.

"I remember one day we had two guys call in with 8630s and they said, 'Yeah, that dirty son of a gun knocked a rod out and the rod came out of the side of the block and knocked the starter right out on the fuel tank.' My employee that was working for me said, 'That's not all. I had two guys within an hour call with the same problem.'"

Kinzenbaw believes the cause of the issue was that the design used the engine as a stressed member of the frame. This put pressure on the engine block. "It was the way they hung 10,000 pounds of front axle onto that engine block," Kinzenbaw said.

He believes the stress the design placed on the engine block, combined with the changes in metallurgy from the block heating up to operating temperature and cooling down, would twist the engine block, which would in turn push engine cylinder sleeves out of place.

The factory fix, in his opinion, was not adequate. "Deere [was] pulling the head off, raising the sleeve up, and slipping a shim around [the] sleeve," he said. "If you have a foundation crushing under your building, you don't jack it up and put planks between it just to make the building the right height. It's going to settle some more."

"I knew that if you went to a salvage yard and found anything from the 1920s all the way through, you'd always find that the original engine block was in those old tractors," Kinzenbaw said. "If you had a John Deere four-wheel drive and it was in the salvage yard, half of the time they had a different engine in them."

The Kinzenbaw team eventually decided there was an opportunity. "We decided we should get back in and repower those four-wheel drives," Kinzenbaw said.

The starting point was to install a new engine—typically a Cummins 855—and construct a front axle frame to take the stress off the engine. The Cummins 855 is one of the outstanding engines of the era, a turbocharged inline six-cylinder diesel that had rock solid reliability and could be tuned to produce more than 500 horsepower.

With a high-powered engine in place, the next issue was to take some stress off the engine. To do that, they turned to history as a source of inspiration.

"You remember the old Olivers had a cast iron frame and the engine set in the frame?" Kinzenbaw said, noting a popular vintage design that used a "bathtub" frame nestled below and around the engine. For the big 1970s and 1980s Deeres, the Kinze engineering team used the same trick, constructing a bathtub frame that surrounded the engine.

"We set the engine in that tub on rubber pads and used a short driveline at the back of the engine and the front end of the tractor," Kinzenbaw said. The machine was longer by about a foot, but the modification was designed to maintain the stock wheelbase.

The Kinzenbaw team also improved access to the engine. "It would take you two men and a boy half a day to take the hood off and get to the engine [on the stock machine]," Kinzenbaw said. His repower included the addition of a tilt-up hood

Kinze continued to offer similar upgrades for the 40 series and 50 series high-horsepower, four-wheel-drive machines.

"The 8850 also was a dinosaur, so they had some serious problems with their 8630s from 1976 clear through the 8850s," Kinzenbaw said. "They had a lot of engine problems."

Kinze Power Products built about six hundred John Deere four-wheel-drive repowers between 1988 and 2007, most of them model 8630s, 8640s, 8650s, and 8850s. It improved its processes along the way, with early repowers known as Series I machines, and the improved version referred to as Series II.

The repower business dropped off for Kinze Power Products once John Deere improved its high-horsepower four-wheel-drive.

"When they came out with the 9000 series, they had a much, much better engine," Kinzenbaw said. "We got on to bigger and better things. We just stayed with the planters and never looked back, I guess you'd say."

THE 40 SERIES

By Ryan Roossinck

MODEL 4440

The "Iron Horses" models brought more horsepower, a refined Sound-Gard cab, as well as higher-capacity radiators, water pumps, and alternators to the line.

John Deere

The 30 series tractors laid a very solid foundation for the 40 series tractors. The Sound-Gard body from the Generation II tractors set new standards for overall operator comfort, and really revolutionized the whole concept of a tractor for America's farmers. With a relatively quiet cab that sealed out the noise and dust, it was a lot easier to handle a full day in the field.

Now it was time to up the ante.

John Deere's 40 series, dubbed the Iron Horses, added a number of innovations to further improve the Sound-Gard body. Better seals and thicker sound deadening foam cut the noise and dust down even more, making the cab an even quieter place to live. Additionally, the introduction of the Personal Posture seat with hydraulic suspension greatly improved the ride and further reduced the vibration.

The biggest improvement to the tractor was its heartbeat. The 404 had been a rock-solid performer for

1979 MODEL 4440 HFWD

Unfortunately, the hydraulic front-wheel-assist system on the 40 series was not much different than the 30 series version. *Renner Collection / Lee Klancher*

close to twenty years, but with larger implements than ever, it was beginning to show its age. Interestingly enough, though, the replacement didn't originate in Waterloo; it came from a dealership in Mexico, Missouri. Eddie Sydenstricker and his service manager Van Botkins campaigned a 4020 pulling tractor called the Cajun Queen. When they needed more power, they built a highly modified 404. Deere's engineers were amazed at the power that it made, and shortly thereafter, the Cajun Queen's heartbeat became the prototype for the 466—a motor that would power tractors for the next sixteen years!

1980 MODEL 4840

Fans of high-horsepower, two-wheel-drives were treated to the 180-horsepower 4840, which came with a standard Power Shift transmission and the updated Sound-Gard cab. *Renner Collection / Lee Klancher*

1979 8640

The big four-wheel-drive model had the same 275-horsepower engine as the 8630. *Machinery Pete*

1980 MODEL 4840

Renner Collection /

Lee Klancher

40 SERIES DATA

Model	Type	Model Years	Notes	HP	Nebraska Test #
840	Utility	1976–1986		40	-
940	Utility	1980–1986		43	-
1040	Utility	1980–1987		53	-
1140	Utility	1980–1987		55	-
1640	Utility	1979–1987		62	-
1840	Utility	1979–1982		60	-
2040S	Utility	1981–1987		75	-
2040	Utility	1976–1982		41	1191
2140	Utility	1980–1987		82	-
2240	Utility	1976–1982	2240-O (Orchard).	50	1192
2440	Utility	1976–1982		60	1085
2640	Utility	1976–1983		70	1157
2840	Utility	1977–1979		80	1249
2940	Utility	1980–1982		81	1351
3040	Utility	1980–1987		88.5	-
3140	Utility	1980–1987		97	1469
3640	Utility	1984–1987		97	-
4040	Row-Crop	1977–1984		90	1267
4240	Row-Crop	1978–1982		110	1266
4440	Row-Crop	1978–1982		130	1265
4640	Row-Crop	1978–1982		156	1264
4840	Row-Crop	1978–1982		180	1263
8440	4WD	1979–1982		180	1323
8640	4WD	1979–1982		228	1324
4040S	Row-Crop	1981–1984		90	-
4240S	Row-Crop	1981–1984		110	-

William Crookes was ten years into a promising career when he began working for the prestigious Henry Dreyfuss Associates (HDA) as a young industrial designer, and he ended up spending his entire career working for the firm. One of his primary accounts was John Deere, and his first major project was a redesign of the Sound-Gard body created by a team consisting of William FH Purcell, Jim Conner, and Chuck Pelly.

Crookes recalls that the initial design of the Sound-Gard was complex, particularly designing the entry door to work with the limited space available next to the roll-over protection system (ROPS).

Interestingly, he added that the soundproof cab actually created a host of unexpected problems to solve. "The seat suspension and seat that they were using on the open station tractors was a very noisy, sloppy system," Crookes said. With the noise of the engine and other external sounds muted by the cab, the seat noise was nearly intolerable.

In the early days of ROPS, there were very few standards. As a result, the solution to the seat problem was less than ideal.

"So a very good friend of mine, Terry Woods, a gifted engineer, attached the seat suspension to the rear ROPS cross member. And when they hit the ROPS from the rear, the seat suspension went with it and they passed that ROPS test. The certifying body told the Deere engineers never, ever, ever, ever do that again. Because the whole idea of providing protection was to keep a zone that was free of any objects that might invade it," Crookes said.

Crookes was assigned to fixing this situation and designing a new seat suspension system. This simple problem turned out to be a major engineering issue to truly fix. The issue was not simply with the seat suspension—it was also a matter of positioning.

The seat and cab had to be positioned roughly directly above the rear wheels for the operator to be able to lean back and see the drawbar and the hitch. This was mission critical positioning of course . . . and Crookes discovered it caused a significant problem. "The location of the cab was just above the rear axle center line and that gave a very difficult ride because the flotation in the tires are effectively a spring."

Any bounce in the tires was transferred directly to the operator. To fix this, the cab would need to move forward, which was a massive engineering change and required creative solutions so the operator could see the hitch.

Another challenge Crookes faced was adapting the seat and controls to accommodate a wider range of body sizes.

"We expanded the recommendations for sizes of operators not to be 90 percent of the population, but to be 95 percent of the population at that time," Crookes said.

Henry Dreyfuss did a tremendous amount of work related to adapting the machine to fit men, a science known as Human Factors Engineering. HDA did work on this for the U.S. military, and developed extensive drawings and measurements showing the ranges of sizes designers and engineers needed to accommodate when creating new machines.

CAB LAYOUT

This wooden mockup was being used by HDA in August 1975.

Henry Dreyfuss Archive, Cooper Hewitt, Smithsonian Design Museum

"For example, if you were backing the tractor up to attach the implement to the drawbar, you had to put your foot on the brake. And you had to put your foot on the brake in the depressed position. And if it were your son, you would want to make sure that he is capable of doing that. So . . . you had to construct the envelope to take into consideration . . . the displacements, the forces . . . for those smaller than average people."

According to Crookes, the tilting and telescoping wheel on the Sound-Gard came about both to accommodate a wide range of operator sizes, and to allow adequate entry space in the tight space between the cowl and the left-hand ROPS structure.

Another issue came about because of the Sound-Gard cab. "Once you closed the cab in, there's a lot of airborne noise," Crookes said. The noise was always there with open station tractors—you just weren't as aware of it. The cab exposed you to noise from the engine, vibration, and from the transmission.

"And the transmission noise could only be addressed seriously if you redesign and test the driveline, hydraulics, and air conditioning," Crookes said. "That's a massive retooling."

In addition, some of the model line used the engine as a structural member, which added more development time and money to a transmission redesign.

"[If] you see a solution, such as a lead blanket that goes over the transmission in a cab, it's to absorb and reduce the noise. Because they didn't want to redesign the transmission."

The vibration issue would finally be addressed in the late 1980s. Engineer Richard Treichel recalls being on the project. "Cab vibration had been an ongoing problem since the Sound-Gard cab was introduced on 1972 tractors," he said. After a long search, he and his group pinned the cause on harmonic motion and were able to eliminate the problem by, among other things, changing the angle of the hydraulic pump.

"You know, things are so intertwined," Crookes said. Fixing relatively minor issues, such as vibration and noise, can require major changes, such as transmission redesigns and relocating the cab. Good tractor design is a complex puzzle that, at times, requires millions of dollars and decades of work to solve.

GENERATION II HOOD

The Generation II tractors employed a brand-new technique for making hoods. The hoods were stretch-formed by a machine that grasped the metal on the left and right hand sides and pulled it down over a form. After the metal was formed, a separate operation came in from the side and pushed in the recess for the left and right hand stripe setting edges. Another stamping created a recess for mounting the cast John Deere logo in the front of the fuel tank.

The new front end was a featured design on the Generation II tractors. Crookes remembers working on the new nose with Jim Conner, a long-time HDA industrial designer and partner who had been creating Deere machinery

FITTING MACHINES TO HUMANS

HDA logged countless hours developing the Deere controls and cab layout to accommodate a wide range of body types and sizes. *Henry Dreyfuss Archive, Cooper Hewitt, Smithsonian Design Museum*

designs for decades. "Generation II had a [new] nose on it," Crookes said wryly, adding that Conner dubbed the new design initiative the "Pick Your Nose Program."

"This was typical of Jim Conner, who was very bright and came up with these things often," Crookes said.

Crookes also recalled that one of the strengths of Deere's design at that point in history was their close working relationship that existed between HDA and Deere. He said that John Deere CEO Bill Hewitt was closely involved in and heavily influenced the design decisions, and he carefully reviewed their rationale and evolution.

"Bill Hewett was very, very interested in appearance and design," Crookes said. "The [HDA] office and its interrelationship was really enhanced by the connection to the very top of this organization."

He said that the industrial design process was integrated into the creation of new machines, their manufacturing, and even followed with critiques after the product was introduced in the field.

"The Deere people had annual product review meetings. Bill Hewitt insisted that these meetings take place, and he critiqued the presented material but was not an active member of the meetings," Crookes said. These reviews included the key Deere leaders, program managers, and engineers, as well as the key staff from HDA. "Hewitt would typically ask if, as of that date, HDA recommendations had been met and whether this met their approval," Crookes said. "There was a closing of the loop."

This meant that Deere's engineering, marketing, advertising, and manufacturing teams were expected to follow through with HDA recommendations and would have to answer to Hewitt if they did not. Crookes added there was a lot of give and take in the relationship, and that Hewitt would also on occasion consult with HDA about the performance of Deere engineers who were up for promotion.

"So long and short of it is, I respected the way the management was conducted between Deere and HDA. And I had some reference to that because of other places that I had worked."

Crookes said that those relationships carried down to engineering and manufacturing. "What's very important was not only the relationship with Moline, but also the good relationship with each of the managers of engineering.

"In Waterloo, there was a manager of engineering, Mike Mack, who was very cognizant of the HDA-Moline relationship. But he also had a great relationship with us because he was implementing what we were recommending and was generally in accord with our activities. So [the relationships with] all of the engineering managers were . . . very much like the interrelationship between Hewitt and us."

Bill Purcell had a John Deere tractor on his ranch, and a dangerous incident inspired change.

Early Deere tractors featured a hand throttle and a spark advance lever mounted beneath, and centered on, the steering column. To increase engine speed, you moved the hand throttle backward. This mechanism was developed partly due to the connection to the injector

pump, and the direction of the motion was due to conventions in the auto industry at the time.

While at his ranch in California, Purcell pulled his tractor up to his barn. "His daughter was standing in front of the tractor, and he pulled the hand throttle backward thinking that it was shutting down," Crookes said.

"He pulled on the hand throttle, the tractor leaped forward, and hit the barn door, stalling the tractor," Crookes said. "She was in the only place that would have been possible for her to survive, in between the fuel tank and the front wheel of the extended axle."

Terrified by the near miss, Purcell insisted that the throttle move to the right on Generation II tractors. This turned out to be an expensive fix due to the position of the ROPS and the steering mechanism.

The relocation ultimately was applied to more than just John Deere tractors, as the throttle on the right positioning became an industry standard. Deere and HDA would work together to create additional industry standards—including the universal symbols that indicate the throttle, brakes, power take-off, hydraulics, and other features on tractors and agricultural equipment.

"[We did a lot to] make them more understandable and in keeping with good human factors," Crookes said.

STANDARDIZED SYMBOLS

HDA developed symbols for vehicles and implements controls that would eventually be used industry wide. *Henry Dreyfuss Archive, Cooper Hewitt, Smithsonian Design Museum*

Hunting Experimentals in Yuma, Arizona

Dale Johnson recalls a fateful phone call that sent him packing his things for a wild goose chase in Arizona box canyons. The year was 1979 or 1980, and Johnson was a John Deere territory manager based in Yuma, Arizona.

This was fertile territory for machine development, as the temperate weather is ideal for testing when the middle of the country is buried in snow. Through much of the twentieth century, new equipment has been moved as quietly as possible to this part of the country and tested as discreetly as possible. To this day, the major manufacturers have proving grounds in the region.

Since the earliest manufacturers tested new equipment in the arid climate of the region, skullduggery went on with engineers and others sneaking around during the dead of night to peak under tarps, crawl under machines, take metallurgy samples, and even partially disassemble equipment.

When Johnson picked up his phone that day, he received a tip from a colleague at the Deere factory in Des Moines, Iowa, that builds cotton pickers. The Deere plant was about to release a new four-row cotton picker, the 9940, and they had a tip to pass on.

"Rumor has it that there is an International four-row experimental cotton picker working around out there some place in southwest Arizona or southeast California," the voice on the phone said. "Do you suppose you could track it down?"

Johnson had a hunch he could.

"I knew who to call, my John Deere dealer in Parker, Arizona, Bill West. Bill was a grand old gentleman and knew everybody in the countryside."

Parker was right on the California border and, as Johnson predicted, West had a hunch how to get a look at the competition's new machine.

"You know, I think I got an idea," West said. "If they're going to run something, there's a real loyal IH guy and

he's in the proverbial box canyon. There's only one way in and one way out. We've gone hunting there."

The canyon was just across the California border, a few hours north of Yuma. Johnson didn't waste any time. "I drove up there to Parker." He was joined by Bill Weber, a salesman for West, and the pair loaded into an ordinary pickup truck. "We threw a couple of shotguns in there and I had my camera."

When they arrived in the box canyon, which was near the Colorado River, they spotted a big red machine picking cotton in the distance. They decided to get closer to get a good look at it, which was when the cavalry arrived.

"The next thing we saw was two pickups come roaring down our way. Dust was flying!" Johnson said. The two trucks skidded to a stop, blocking Weber and Johnson's vehicle in the front and rear.

"What are you guys doing here?" the IH man asked.

"Well, we're scouting out hunting for bird hunting this fall," Bill said.

"What are you doing with that camera?"

"Oh, we like to take pictures too."

"Bullshit! You guys are from John Deere, aren't you? Get the hell out of here!"

The two men high-tailed it out of there and went back to the dealership to consult with West.

The mission wasn't over, as West had an idea.

He owned a four-wheel drive dune buggy with a big engine that he thought was just right for the job.

"You know what," he said to Bill Weber, his salesman, "Why don't you take Dale out in the dune buggy and you can just go through the desert and get on top of that canyon, and I'll bet you can find a way to slither down the canyon wall and get over there."

So Weber and Johnson beat their way off-road to the top of the canyon.

"We found kind of a drainage, a crevice in the canyon wall, and so we slithered down that canyon wall and got to the bottom to the edge of the cotton field," Johnson said.

"Then we low crawled, you know, we used some of our old army skills and we low crawled between the cotton

EXPERIMENTAL COMBINE

This is the John Deere 370-M 1 XH71 experimental combine model developed for Harvester Works. Image dated November 1982.

Henry Dreyfuss Archive, Cooper Hewitt, Smithsonian Design Museum

rows and towards the cotton picker and it happened to be just right at noon time and so they had stopped the cotton picker under a clump of cottonwood trees along the irrigation canal."

The IH crew was sitting under trees having lunch, and Johnson was able to sneak over and photograph the experimental cotton picker. He climbed on top of the machine, and inside it as well.

"We low crawled back out of the cotton rows and climbed back up out of the canyon, and it was a lot easier going down than it was going up, I remember that."

On the rim, they discovered a deer or elk skeleton.

"We propped up the head and I took some pictures of that too and so we called it the skull espionage," Johnson said.

The photos were sent to Deere headquarters, and word got back that they were passed around to a large number of employees there, eventually finding their way to the CEOs office.

Deere introduced the 9940 four-row cotton picker, and International put out its Model 1400 four-row picker. Johnson doesn't recall who was first to market; the records on this are hard to find at best. No matter— the two companies were at it hammer and tong for this market.

"That was pretty tough competition back in those days and so we were both trying to gain a foothold. We had probably 60 percent market share on cotton pickers, on the two-row cotton picker and so they had the rest. There were only two cotton picker manufacturers, Deere and IH. They had probably about 40 percent and we were wanting to keep growing in that market share and so it was pretty intense," Johnson said.

The recession of the early 1980s had profound effects on the agricultural machinery industry. In the fast-paced growth of agriculture in the 1970s, farmers took opportunities to leverage themselves with debt for both land and equipment, given what seemed to be a halcyon future. In the process, the farm machinery industry did well. Deere, International Harvester, and Massey Ferguson continued as giants of the industry, with Allis-Chalmers, Ford, J. I. Case, and New Holland taking significant roles, too. Two companies captured an increasing share in the four-wheel drive tractor market—the Vancouver, B.C., firm of Versatile Manufacturing Company, and the Fargo, North Dakota, firm of Steiger Tractor Company (the latter with a minority holding by International Harvester from 1975). The mid-1970s also brought a new player on the North American scene with the entry from Japan of Kubota, Ltd.

When the farmers' hopes were dashed in the abrupt turnaround, one of the inevitable results was a "make do" mentality on equipment purchases. In the face of a deadly combination of high interest rates and falling farm prices, many potential agricultural machinery purchasers just opted out. The farm equipment industry was left in disarray. Business sagged precipitously, and just about every farm equipment manufacturer, large or small, was faced with cutbacks in schedules, stretchouts in capital spending plans, and, frequently, reductions in staff.

Some manufacturers fared relatively much worse than others. Massey Ferguson experienced its own set of problems even before the onslaught of the downturn; in 1978 it lost a startling $262 million. After a small profit in 1979, the company again lost heavily in 1980, more than $225 million. The firm was saved from bankruptcy only by a massive financial transaction involving a group of some 250 banks and insurance companies from around the world, together with the Canadian and British governments. Even this $730 million (CDN) rescue package could not stem the tide of red ink for Massey Ferguson; the company lost $194.8 million in 1981 and losses continued in 1982.

International Harvester had only modest profits in 1976–78, and a much better performance in 1979. In 1980 the combination of the business sag, an abortive strike, and certain internal management missteps brought a colossal loss of $397 million. In December of that year, Harvester, too, had need for a rescue plan, in this case involving some two hundred banks and insurance companies, with a total package of $4.15 billion. Nineteen eighty-one brought a further loss of $393 million; in May 1982 Archie R. McCardell, the chief executive officer from 1977, resigned. The continuing saga of bad news at these two business giants is well known; their difficulties were not just an industry story, for the general press throughout the country widely chronicled the two firms' efforts to forestall bankruptcy.

But there were also many other changes upward and downward in the fortunes of individual companies. Market shares were indeed shifting in a way hardly imaginable in prior years, and some famous names were almost gone altogether from the scene. The White Motor Corporation (the successor combination of Minneapolis-Moline and Oliver) went bankrupt in 1980; its farm equipment division was sold to the TIC Investment Corporation, a Dallas firm. Operations for White continued, but at a low volume with substantial losses. Others, and this included preeminently Deere, picked up market share and relative strength in this period, even while they themselves struggled to maintain equilibrium in the face of very depressing sales outlooks.

The two charts shown here picture in a dramatic way the market share position of Deere in agricultural equipment. North American farm equipment sales graphically show Deere's strong position in the North American market, recording the company's nearly 30-percent piece of all farm equipment sales.

The company's reach worldwide is pictured in Farm Equipment Sales Worldwide; its 17 percent market share is the only one in double figures for 1980. Deere's overseas success has been greater in combines than in tractors, and there are only a few markets in which the company has a "substantial" market share of 20 percent or more— Spain (tractors, combines), Australia (tractors), and Brazil (combines). Deere's progress has been painfully slow in the two key markets of France and Germany.

FARM EQUIPMENT SALES IN NORTH AMERICA: MARKET SHARE AND GROWTH RATES, 1970-80

Company	Year			Annual Growth Rate		
	1970	1975	1980	1970–75	1975–80	1970–80
Deere	26%	25%	29%	21%	12%	16%
International Harvester	18	17	14	21	5	12
Massey Ferguson	8	8	7	20	7	13
Fiat	NA	...	3	NA	NM	NA
Hesston	1	2	NA	34	NM	NM
Sperry New Holland	4	3	5	16	18	17
Ford	5	4	4	18	9	13
J. I. Case	4	6	6	29	8	18
Allis-Chalmers	6	6	5	22	7	14
Versatile	1	1	2	91	6	42
Steiger	...	1	1	91	6	42
Claas	NA	...	...	NA	NM	NA
Deutz	NA	...	1	NA	79	NA
Kubota	NA	NA	NA	NA	NA	NA
Total Industry Sales	$3[b]	$8[b]	$12[b]	22%	9%	15%

NA – Not Available; NM – Not Meaningful
[b]$000,000,000

FARM EQUIPMENT SALES WORLDWIDE: MARKET SHARE AND GROWTH RATES, 1970-80

Company	Year			Annual Growth Rate		
	1970	1975	1980	1970–75	1975–80	1970–80
Deere	15%	16%	17%	23%	12%	17%
International Harvester	12	14	9	25	4	14
Massey Ferguson	10	12	9	24	5	14
Fiat	NA	4	5	NA	19	NA
Hesston	1	1	NA	38	NA	NA
Sperry New Holland	4	3	4	19	14	17
Ford	7	5	4	14	6	10
J. I. Case	2	4	4	32	10	21
Allis-Chalmers	3	3	3	21	7	14
Versatile	...	1	1	34	20	27
Steiger	...	1	1	94	8	45
Claas	NA	2	2	NA	11	NA
Deutz	NA	2	2	NA	18	NA
Kubota	NA	4	4	NA	10	NA
Total Industry Sales	$6[a]	$16[a]	$27[a]	21%	12%	16%

[a]$000,000,000

THE 50 SERIES

By Ryan Roossinck

Deere introduced some major innovations to farmers with the 50 series. This wasn't a simple cosmetic enhancement and a small bump in horsepower. The combination of a gear driven mechanical 4WD system, a fifteen-speed Power Shift transmission, and caster action made field work more efficient than ever before!

Prior to 1983, most of Deere's row-crop 4WD systems were run off the hydraulic pump, and they really weren't all that great. They were notoriously unreliable, didn't like to work when it was cold, and they were expensive to maintain. With the introduction of the 50-series tractors, the company implemented a mechanical, gear-driven system. It was a lot more reliable, less expensive to maintain, and unlike the hydraulic system, was designed for full-time use.

The fifteen-speed Power Shift transmission was, in most cases, better for field work. It eliminated a lot of the "in between" issues of the previous eight-speed Power Shift, offering more gears to effectively use the horsepower. Furthermore, because the gearing wasn't spaced so far apart, shifts were smoother! (At the end of the day, it may have been farm kids who saw the most benefit from this—they banged their heads against the back window a lot less frequently when riding with Dad!)

Lastly, caster action. Caster action was a system that tilted the kingpin on the front axle for sharper turns. It was faster and more efficient because it used less fuel. It wasn't perfect and had to be manually activated, but it definitely saved farmers time and money.

The 50 series lineup was the largest product line of new tractors in the company's history. Between 1981 and 1986, Deere launched twenty-two new tractors, nineteen of which were sold in the U.S. market. It was a bold move, given the state of the farm economy at the time.

EARLY 50 SERIES PROTOTYPE DESIGN

This model was built by HDA in June 1977 and is labeled "SGB with side tanks." *Henry Dreyfuss Archive, Cooper Hewitt, Smithsonian Design Museum*

DESIGN PROTOTYPE

Another HDA model, also dated February 14, 1977, is a larger, longer machine. *Henry Dreyfuss Archive, Cooper Hewitt, Smithsonian Design Museum*

DESIGN PROTOTYPE

Another HDA model, this one dated February 14, 1977, is a variation of the model above. *Henry Dreyfuss Archive, Cooper Hewitt, Smithsonian Design Museum*

1983 4250 HI-CROP FIRST PRODUCTION

Deere's new for 1983 50 series represented a major update to its farm tractor line. The fifteen-speed Power Shift was the star of the show, along with increased horsepower and features. *Keller Collection / Lee Klancher*

MODEL 4450

The 4450 offered 140 horsepower and the new transmission. This 4450 was photographed May 11, 1987.

Henry Dreyfuss Archive, Cooper Hewitt, Smithsonian Design Museum

1988 MODEL 4450 MFWD

The new mechanical front-wheel-drive system on the 50 series tractors was more powerful and reliable than the hydraulic version on past models, and is highly regarded today.

Machinery Pete

1983 MODEL 4650

The 165-horsepower 4650 and 190-horsepower 4850 offered enough power and traction to appeal to former buyers of the larger four-wheel-drive models. *Machinery Pete*

50 SERIES DATA (see page 304 for 8X50 series data)

Model	Type	Model Years	Notes	HP	Nebraska Tractor Test #
1350	Utility	1986–1994		37	-
1550	Utility	1987–1994		43	-
1750	Utility	1987–1994		50	-
1850	Utility	1986–1994		55	-
1950	Utility	1988–1994		61	-
2150	Utility	1983–1986		-	1470
2350	Utility	1983–1986		55	-
2450	Utility	1987–1994		70	-
2550	Utility	1983–1986		65	-
2650	Utility	1987–1994		78	-
2750	Utility	1983–1986	2750 Hi-Crop, 2750 Low-Crop (Orchard).	75	-
2850	Utility	1986–1994		86	-
2950	Utility	1983–1988		85	-
3050	Utility	1986–1993		84	-
3150	Utility	1985–1987		-	1472
3350	Utility	1986–1993		-	-
3055	Utility	1992–1993		94	-
3650	Utility	1986–1993		114	-
4050	Row-Crop	1983–1988	4050 Hi-Crop.	100	1457
4250	Row-Crop	1983–1988	4250 Hi-Crop.	120	1458
4450	Row-Crop	1983–1988		140	1459
4650	Row-Crop	1983–1988		165	1460
4850	Row-Crop	1983–1988		193	1461

In 1983, Deere & Company stepped into the world of rotary engines, commonly known as Wankel engines, invented in the late 1940s by German engineer Felix Wankel. The company purchased patent rights and technology related to rotary engines from Curtiss-Wright, a company formed in 1929 with the merger of companies founded by the Wright brothers and Glenn Curtiss, who is considered the father of naval aviation.

The purchase gave Deere exclusive North American rights and patents to the Wankel engine, as well as know-how, experimental engines, and components. The best-known production vehicle powered by a Wankel engine was Mazda's RX-7 automobile beginning in 1979. Deere engineers had determined the existing rotary engine platform was sound, so incorporating this technology with Deere & Company's extensive research, design, and manufacturing facilities promised to bring a reliable product to the marketplace.

The company's effort considered the rotary engine's compactness, eliminating upwards of 50 percent of the

ROTARY ENGINE PROTOTYPE

This nicely finished prototype was developed when John Deere purchased the rights to the Wankel rotary engine from Curtiss-Wright. Development was done in a New Jersey facility that was part of the purchase. The 580 engine used two rotors. *Mannisto Collection*

bulk of reciprocating diesels, multi-fuel capabilities and economies, design simplicity, and optimum manufacturing cost.

Jon Lauter, Dick Hurban, and John Mannisto were all freshly minted engineers when they joined the new team that was a mix of both Deere & Company employees and those from Curtiss-Wright, which had a research facility in New Jersey. That interaction proved both fruitful and frustrating at times as each business entity entered manufacturing worlds neither had experienced in the past. This was a brave new world in engine technology for Deere while Curtiss-Wright, which was used to major contractual relationships, was teamed with an ambitious and energetic group of designers and engineers who wanted to test the limits of this innovation.

Jon Lauter was one of two engineers hired as the first non-Deere, non-Curtiss-Wright employees for the Rotary Engine Division (REDIV) in April 1985. Signing on as a Development Engineer, Lauter had just graduated from Hofstra University with an Engineering Science degree.

"I had several years' experience as an auto mechanic," Lauter explained, "including experience with Mazda's rotary engines." Little did Lauter realize how that first position would impact his working career as he moved even further into rotary engine development after Deere shut down the division in 1991, selling it to Rotary Power International, Inc. (RPI), which went out of business in the early 2000s, and the Wood-Ridge, New Jersey, site has been abandoned.

Lauter worked on combustion sealing issues and general engine performance and durability during his years at Deere and left the company upon the demise of REDIV, going to work for Cummins Engine Co. in Columbus, Indiana, as a technical specialist. In 1994, however, he returned to the rotary engineering world, taking a position with RPI in New Jersey, assigned as a project engineer for a small .4-liter (about 24 cubic inches) engine being developed for unmanned air vehicle (UAV) applications. He also later managed the engineering

effort for a marine engine product based on the Mazda RX-7 rotary engine block.

John Mannisto also joined the REDIV team in 1985, graduating from Rutgers University in 1983 with a degree in mechanical engineering and then working for Lockheed Corp. for 2.5 years. He described that first job interview in one of the small original buildings, crowded with engine parts and drafting boards. "I was hooked!" Working for Deere & Company, taking on an exciting new project with a startup feel and given plenty of responsibility, Mannisto admits those years have been some of the best of his career.

"I was hired to help develop simulation capability," Mannisto described. "This was the simulation of thermal and structural behavior of engines on the computer, something I had been pursuing in my master's degree work." He said Deere & Company paid for his master's degree work in 1987, and he went on to engine development after three years in the simulation group.

ENGINE DEVELOPMENT

The rotary engine piston design can be seen here—the center piece rolls around as the engine fires. Deere engineer John Mannisto is shown working on the computer. The most famous rotary engine vehicle was Mazda's RX-7. *Mannisto Collection*

"There was always a lot of enthusiasm around our work," he recalled. "I think we all knew this was pretty special. There was a lot of work and a lot of pressure, but I never heard anyone complain. There were times when I slept in the test cells!"

Mannisto's time in the engine build and testing group was a good match, from his view, for both him and the team. "I got to learn a lot about engine testing, and I brought some computer methods to the data acquisition. You must remember this was a very new thing back then. The PC was only a few years old!"

Dick Hurban, who still works for Deere & Company at the time of this writing, also graduated from Rutgers University in 1983 with degrees in Agricultural Science and Agricultural Engineering before completing his master's in 1986 in Agricultural Engineering with an emphasis on the flow of fibrous materials.

He first saw an advertisement for John Deere Rotary Engines in January 1986 and thought it was a dream come true. "As an Agricultural Engineer I was of course attracted to companies like Deere, Caterpillar, or IH Case," he said. "I had always been interested in engines and the rotary Wankel engine."

He described how one of his college professors owned a Mazda RX-2 that had an engine replacement and used the replaced engine in some of his classroom work. "I became a big fan of the Mazda RX-7 when my brother purchased one. They were really fun to drive!"

Hurban was hired in May 1986 after completing his master's degree. "My first role was as the Development Engineer in charge of building and running our two-rotor 580 engines for two years. I then became a design

engineer for the three-rotor 3174 engine, which was being funded primarily by the U.S. Marine Corps," he said. "The engine was doomed, however, due to some critical crankshaft integrity issues and eventually we found the engines were not viable."

He also echoed the enthusiasm of both Lauter and Mannisto for the project. "I think 90 percent of the team were true believers that we [Deere & Company] could make the engines work and be viable. Although the three-rotor was not feasible, many believed that power packs consisting of multiple two-rotor units could be used, perhaps even in larger engines."

"My five years at REDIV were certainly the most fun I have had in my thirty-five years with Deere," Hurban said. "Every day brought new problems and opportunities. We had fellow engineers who had worked on the first helicopters, the first jet engines, and even designed the first nuclear reactors for Trident submarines. They were willing to teach, mentor, and even learn from new graduates and young engineers."

Mannisto described the working environment during those years as consisting of three "camps." "There were the original Curtiss-Wright employees, many of whom had been on the project for decades and had rotary engines in their blood with a strong commitment to the technology."

The second camp consisted of Deere & Company transfers from the Midwest, most coming from the Waterloo Engine Works. "Many of those people took this project because of the career opportunity to work on a new technology away from the standard diesel engines they were used to working on."

PROTOTYPE GRAPHIC

The development of the line went quite far, with graphics and banners created.

Mannisto Collection

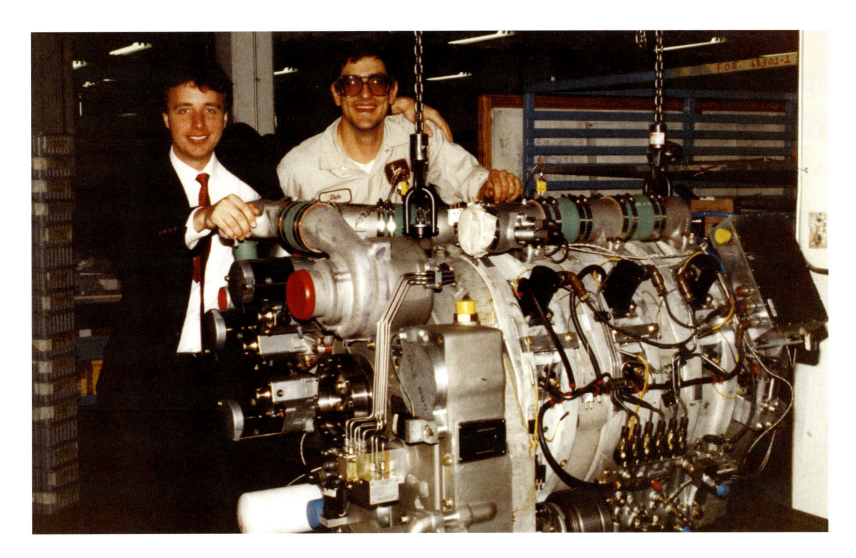

The third group included Mannisto, Lauter, and Hurban, who were all new and had no history with either Deere or Curtiss-Wright. "We were all thrilled to be involved in this project."

Mannisto described the "camps" as having specific differences in how they operated. "The Curtiss-Wright group was used to "shooting the moon" and working with crazy demo applications to attract funding. There were some cool things stored in the complex like a 1966 Mustang with rotary RC2-60 engine, a speed boat, and even a lawnmower."

Hurban recounted how he and another engineer who was somewhat older were getting a prototype engine to a gathering in a room located up several flights of stairs.

"I remember thinking that carrying this engine up those stairs was going to be a problem, but somehow the two of us got the parts to the proper place, assembled it and made the presentation," he said. "I've often thought since how strange that is compared to the remainder of my career where engines are so big and heavy, we would never be faced with carrying one up a flight of stairs."

"The Deere people were more practical, methodical, and cautious with an eye on long-term production of a real engine," he described. "I think that is what attracted the Marine Corps to the new division. They saw the innovative nature of the Curtiss-Wright team tempered by John Deere experience with making production engines. And for the most part that marriage worked."

EARLY PROTOTYPE ENGINE

John Deere engineers John Mannisto (left) and Dick Hurban with the engine developed by the John Deere team. *Mannisto Collection*

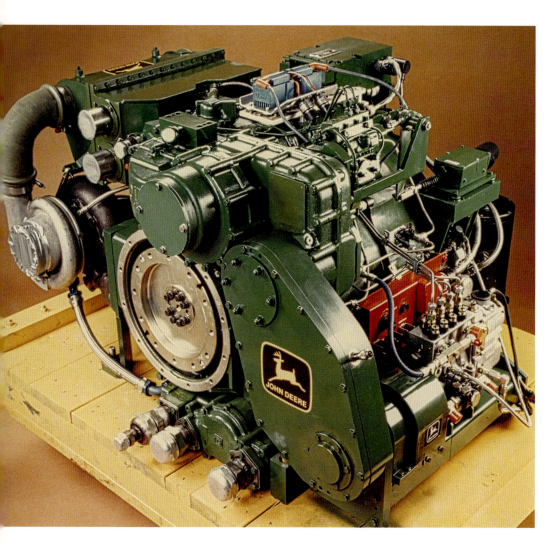

DEERE 3174 THREE-ROTOR ENGINE

This engine was developed for use by the U.S. Marine Corps. The engine never saw production, and the entire John Deere rotary engine program was terminated in spring 1991. *Volgarino Collection*

Meshing ideas and processes was a challenge for all the groups involved and, in some cases, there were people who got pulled into the project through no effort of their own. "During the time REDIV was in operation, Deere core markets were stagnant or contracting," said Lauter. "As REDIV staffing increased, many long-term Deere employees accepted transfers from Iowa or Illinois to New Jersey, rather than face layoff."

"Of course, some did so enthusiastically, but for others there was a reluctance," he explained. "Some brought great skills and expertise while others did not and none of them had any rotary-specific knowledge or experience."

Lauter said as the division grew rapidly there were many missed technical and business targets, which in

many cases were simply set unrealistically. "Unfortunately, a degree of malaise set in, and a sense of failure was expected more often than not."

Hurban echoed some of those thoughts, but said he thought management, for the most part, felt the group needed to "do its best work" and not be frightened by failure. "I felt like this was a way for us to learn."

In the spring of 1991, the president of John Deere Technologies International (JDTI), Edward Wright, visited the New Jersey facility, and Mannisto describes how all the employees crowded into the area where the engines were built. "Ed announced that Deere would be getting out of the rotary engine business." He said many of the group had suspicions during the previous year that the project was not going to continue, "but this simple announcement made it real." Mannisto said many people were visibly shaken and some cried. "It was a tough blow for us."

"Deere had come to see this technology was not going to position the company with a volume market engine, and Deere made its money selling engines, not selling development contracts as Curtiss-Wright had done for decades," he explained. "When we could not meet some of our development targets in performance and reliability, [Deere management] didn't see a path to something they could manufacture and sell."

Hurban thought another key factor was the collapse of the Soviet Union in 1990. "It appeared all military research and development contracts were going to dry up and Deere was struggling to meet Department of Defense performance requirements. Those facts also helped put Deere management on the defensive and the division was sold within months."

Would the division, if it were in operation today, survive the challenges presented back in the '90s? All three agreed that today's technology would have made a huge difference in the development of the engines by increasing efficiency and all-around performance using technologies that have since been brought into play.

Mannisto said ironically one of the markets, a small series of rotary engines for use in flying drones (something that would be science fiction in 1990), was developed successfully by a team at Norton motorcycle. "They had developed a Wankel for motorcycle use but used their expertise to develop the technology for aircraft," he said. "The company, UAV Engines in England, was the result." Visiting the company in 2006 while a consultant, Mannisto said "the place was alive with production and development with lots of military contracts. It was great to see someone succeed with this!"

FINDING SPACE IN NEW YORK

As part of the rotary engine complex, HDA was granted some much needed and appreciated space. This additional space permitted full-size mock-ups, as well as the ability to receive tractors, implements, lawn and garden equipment, and a wide range of John Deere products.

HDA relocated facilities several times. As William Crookes recalls, the space on Fifty-Fifth Street had room for two large drawing boards shoehorned in. "We had two drawing boards that were—without exaggeration—twenty feet long and four feet tall."

These were great for creating large-scale concept drawings, but their service elevators were too small to bring up tractors and their parts. The boards were helpful but were no substitute for walking around life-size machines. And, Crookes added, this was prior to the era when all design was done on a computer.

"Deere thought that they were going to get into the rotary engine business, and they bought a company in Woodridge, New Jersey, and it was called John Deere Rotary Works."

The purchase of the company included a huge building in New Jersey in which Curtiss-Wright had built radial engines for the World War II effort. HDA negotiated with the engineering managers at Rotary Works to get a little bit of the space for John Deere development.

As offered, the space was not ideal. The floors in this building were made of cut hardwood blocks (end grain facing up), much like many typical high school shops.

"These floors had absorbed oil and solvents and so on to where it was impossible to clean or keep clean," Crookes said. "It smelled like oil residue, and it wasn't often level."

Ambience and floor details aside, the space was welcome. "That gave us the opportunity to receive tractors and additional equipment and gave us the opportunity to get on, get off, get under, and so on and so forth," Crookes said. "We built an internal office space within to accommodate Catia computer stations and terminals. Count Curtis (the son of the CFO at Deere, Woody Curtis) was responsible for providing these (including training) for Deere. Ultimately, HDA added ALAIS, CDRS, ICEM, and ProE design softwares to these stations," Crookes said.

When John Deere introduced its new 50 series tractors in fall 1982, it put on a show in the New Orleans Superdome that was nearly as flashy as Deere Day in Dallas, when it introduced the New Generation tractors in 1960.

John Deere territory manager Dale Johnson remembers the day in New Orleans well, both because of the spectacle and one memorable intrusion.

"Deere had rented the New Orleans Superdome for the new tractor intro because it was going to be a huge intro with the new 50 series tractors with the fifteen-speed Power Shift transmission and the mechanical front wheel drive and more horsepower," Johnson recalled.

"They had laser lights and live music and they had dancers and performers and it was smoke and fire," Johnson recalled. With dealers from around the world, the event was quite a show.

Word of the product launch had traveled to the competition, who very badly wanted to get an early look at the new line.

Pulling shenanigans to get a sneak peek at the latest machine of the rival color was nothing new in the ag industry. All sides had personnel out in the field looking specifically for test machines, and stories abound of stunts pulled to get an advance look at the competition, including tearing apart prototypes left unattended, sneaking around at midnight to test metals, and camping out with cameras and telephoto lenses at likely test fields.

For the Superdome 50 series launch, John Deere had posted security at the door and badges were required for entry, but a fast-talking sort with a John Deere cap and shirt just might be able to get inside.

"During the show, one of our security people thought he saw some kind of a reflection on the roof of the Superdome," Johnson said. "They've got a huge structure up there and catwalk with lights and speakers and they hang the big screen from there."

Johnson remembers the security telling him he was headed up to investigate.

DEVELOPMENTAL MODEL

This R76R developmental model is dated August 30, 1982. The model is labeled the JD 365—all the Deere developmental machines had similar numbers.
Henry Dreyfuss Archive, Cooper Hewitt, Smithsonian Design Museum

The reflection on the Superdome's roof was from International Harvester salesman Bud Youle, who had talked his way in, and scampered up to what he thought was a safe spot.

"I had been in [the Superdome] before," Youle said. "I knew there was a bar up [above the field]. I thought if I could get up to that bar, and I could sit and listen, that's all I needed to do.

"I went to the elevator. I went straight up to the top . . . and I sat down up there in the dark."

He was able to sit there for a good while, listening to the presentations of new machines. As time wore on, he recalls thinking he'd better move. As they say in the military, never stay in one spot too long.

"I got myself a Budweiser beer, and got in a different place," Youle explained, "but I could still hear them. Pretty soon, I heard the elevator come up."

That was Dale Johnson's security guard. "He went up there and low and behold, he came upon this guy that was taking pictures," Johnson said. "He called down for some help and then ended up grabbing a hold of him, so they got him down off the catwalk and got him into their security room and then started asking him a whole bunch of questions."

As an aside, as Youle told the story, he was in a bar. Johnson's story suggests Youle was on a catwalk. While there was a bar high in the Superdome so it's possible Youle is correct, Youle loves to add zest to his stories. Given his storytelling style, the smart money is Johnson remembers correctly and Youle was on a catwalk.

Both parties recall Youle coughing up his identity pretty much immediately.

"My name is Bud Youle, Harvester Company," Youle remembered saying. "I was down here, and I saw what was going on."

As Johnson tells it, the Deere folks speed marched Youle back down to a main office and then placed a series of phone calls trying to figure out what to do with him.

"They had an idea of what they'd like to do with him, but they didn't," Johnson added with a laugh.

They finally were able to reach new Deere CEO Robert Hanson to ask for some help figuring out how to properly dispose of the Spy.

Johnson and Youle both recall Hanson and the new CEO at International Harvester, Louis Mencks, having words to settle the matter.

According to Johnson, Hanson called Mencks and said, "We're trying to figure out what we should do with your guy here. We've caught him taking pictures and we don't know whether we should shoot him or just let him go back to you."

Mencks must have suggested the latter option, as the Deere folks took away Youle's camera and released him to scamper home to Harvester.

Johnson added that the espionage didn't surprise him. "I mean we always knew when IH was going to be introducing new tractors. There's always some speculation, the word gets around, you know they run test tractors, we run test tractors. You put them in farmers' hands, and somebody might spill the beans at a coffee shop, you know?"

"It's not too unusual for those kinds of things to happen," Johnson said. "It's pretty unusual to get caught."

HIGH-HORSEPOWER 50 SERIES

Farming's toughest year in the modern era was 1982. John Deere had historically weathered hard times with extreme conservatism, but the company in the 1980s had been number one in the United States for nearly two decades. In an uncharacteristically aggressive move, the company unveiled a line of new machines in the fall of 1982. The 8450 and 8650 replaced the 40 series machines, and the flagship of the line was the big 8850, with power from a 955-cubic-inch V-8 good for 370 horsepower and an advanced Quad-Range partial Power Shift transmission.

The machine wasn't perfect, but it was worlds better than the 30 and 40 series high-horsepower four-wheel-drives. This solidly built line helped John Deere become the only tractor manufacturer in the United States to show a profit in 1982.

MODEL 8850

Deere followed the market and offered its flagship four-wheel-drive with a 300-horsepower turbocharged and intercooled 955-cubic-inch V-8 diesel engine.

Marcus Pasveer

MODEL 8850

Unlike the 40 and 30 series four-wheel-drive Deeres, the 50 series featured tilt-up hoods. *Marcus Pasveer*

GRILL DESIGN SKETCH

October 1978 sketch from HDA. *Henry Dreyfuss Archive, Cooper Hewitt, Smithsonian Design Museum*

1983 8650 REPOWER

Kinze Manufacturing fitted this Deere with a Cummins engine. *Machinery Pete*

4WD 50 SERIES DATA

Model	Type	Model Years	HP	Nebraska Test #
8450	4WD	1982–1988	187	OECD 1436
8650	4WD	1982–1988	239	OECD 1435
8850	4WD	1982–1988	304	OECD 1434

There was a major update to the Waterloo manufactured tractors planned for introduction in fall of 1982. With the addition of the hi-lo Power Shift and oil cooled clutches to the Quad-Range transmission, there was a need for more speeds in the eight-speed Power Shift transmission. I suggested that we could add a hi-lo planetary to the Power Shift to provide a fifteen-speed transmission. But we had to mount the clutch to the engine and let the pressure supply oil manifold float with the movement of the engine crankshaft. My eighth patent (4,373,622) showed how to provide a floating manifold for a clutch mounted to the engine crankshaft. I was the transmission project manager for the redesign of the Power Shift for the fifteen-speed transmission.

SOLVING POWER HOP

In addition to the transmission upgrades, mechanical front wheel drive was developed for the 50 series tractors. In 1977, I was made Manager of Advanced Drivetrain Engineering. I had a staff of a couple of engineers capable of doing computer programming and analysis and a few design engineers working on advanced transmission projects.

While developing the mechanical front-wheel-drive equipped tractors, a problem called "power hop" was encountered and developed enough concern that I was assigned to a group to study the problem. Power hop is a phenomenon where the tractor starts to bounce like a ball while pulling heavy draft loads in soft, dry soils. This had been a problem with the four-wheel-drive tractors (large drive wheels that are all the same size) since the 8010 tractor was tested in 1958. A model was tested at Purdue University and mathematical modeling was done at the John Deere Technology Center without successful results. I did some mathematical modeling also without success, but I decided to look at the root locus system stability analysis from my Automatic Control Engineering advanced course. Since the problem occurred in soft, dry soil, I decided that I had to develop a model that included tire, traction, and soil characteristics as well as wheel sinkage into the soil and geometry. My stability analysis appeared to correlate well with field test results with changes in tire pressure, difference between bias and radial tires, different soils, and foam-filled tires. This seemed to verify that the problem was a system stability problem. Tests found that the system stability was very sensitive to soil moisture, even deep in the soil. In one case, in irrigated soil, it was found that a field that was slightly dryer would cause instability and an adjacent wetter field would not. It has been said that the best way to control power hop is to have a rain! My work on power hop was stopped when it was found that putting high pressure in the front tires and low pressure in the rear tires solved most of the problems.

8850

To quiet the ComfortGard cab the transmission had to be modified. As design progressed, engineers became more aware that systems are related and simple problems can require expensive development.
Henry Dreyfuss Archive, Cooper Hewitt, Smithsonian Design Museum

55 SERIES

By Ryan Roossinck

In 1989, John Deere released what most green tractor fans regard as the "ultimate" version of the Sound-Gard tractors. The 55 Series was the culmination of nearly twenty years of development, and while the changes were incremental compared to previous versions, they were the final touches needed to turn a good tractor into a great tractor.

All six of the Waterloo-built models now featured a turbocharged version of the venerable 466. Thanks to redesigned intakes and injectors, Deere was able to provide a 5–10 horsepower increase across the lineup as well. The additional horsepower was timely, as implements were more power hungry than ever before. Transmission options were unchanged; why mess with a good thing?

Deere also made further refinements to the caster action system in the MFWD models. In the 50 series tractors, the operator would need to manually engage it as they reached the end of the row. Once engaged, the kingpin in the front axle would tilt, allowing for much tighter turns (18 feet, according to the advertising). In the 55-series, though, caster action engaged automatically—one less thing for the farmer to worry about!

Inside the cab, Deere continued to update materials to make the cab a more comfortable place to be. It also upgraded the dashboard to a fully digital version, scrapping the last of the gauges from the 50 series, and redesigned the window latches for easier operation.

The 55 series sold very well, and thousands of them continue to earn their keep on farms across North America!

55 SERIES

In 1988, John Deere released the 4055 series of tractors, ranging from 117 to 222 horsepower. The revised machines boosted horsepower, with updated drivelines to match. Although this was refinement more than revolution, the new machines were popular. The big news for the year? John Deere topped the sales record set in 1979 with the new line. *John Deere Archives*

1991 MODEL 2955

The Mannheim-built compact and utility tractors were updated for 1987. Several transmissions were available, and some models offered mechanical front-wheel-drive. Orchard and high-clearance versions of some models were offered.
Machinery Pete

1990 MODEL 4055

The entire mid-range line of 4X55 John Deeres used the 466-cubic-inch diesel six-cylinder engine.
Machinery Pete

1990 MODEL 4255

The 4255 was rated for 120 horsepower, and a high-clearance version was offered. *Machinery Pete*

1991 MODEL 4455

The 140-horsepower 4455 could be had with an optional mechanical front wheel drive. The model was built in Saltillo, Mexico, as well as Waterloo, Iowa. *Machinery Pete*

The first ingredient of "using capital for strategic purposes" is to understand the underlying trends in the market, and Deere seems to have done this very well indeed. A major *New York Times* article comparing Deere and International Harvester concluded: "Deere correctly saw the post-World War II trend toward fewer and bigger farms. From this came its strategy: continual reinvestment in product and manufacturing innovations, close attention to costs and quality, and lavish support for its dealer network. As a result, Deere is in the enviable position of being the low-cost producer and the high-quality provider."

The forty-five-year-long decline in the number of U.S. farms was reversed in 1981—there were 8,000 net new units in the 1981 farm count for that year. Officials of the U.S. Department of Agriculture (U.S.D.A.) estimated, however, that most of these were in the small-farm, part-time category likely reflected in the movement away from commercial farming.

The commercial potential of American agriculture seems oriented ever more strongly toward the large farm (defined by the U.S.D.A. as those with gross annual sales of $100,000 or more). By 1981, the largest 1 percent of the farms in the United States was producing about one-fourth of the nation's food; by the year 2000, according to the U.S.D.A., the largest million farms will operate almost all of the nation's farmland. Three-fourths of the farmland will be in the hands of the top 200,000 operations; indeed, the largest 50,000 will produce about two-thirds of all farm output.

Deere analysts consistently emphasized the central importance of this group to the agricultural machinery industry. In the census of agriculture of 1978, those farms with annual sales of $40,000 to $499,999—some 560,000 units—accounted for almost 60 percent of farm products sold and in the process took more than 70 percent of the industry's machinery sales. As the *New York Times* analyst put it, "Deere had the strategy and foresight to see where the market was going . . . and became the market leader in large equipment, which carries the highest margins."

55 SERIES DATA

Model	Type	Model Years	Notes	HP	Nebraska Test #
2155	Utility	1987–1992		45	2024
2355	Utility	1987–1992		55	2025
2355N	Orchard/Vineyard	1987–1994		55	-
2555	Utility	1987–1992		65	-
2755	Utility	1987–1992	Orchard & High-Clearance.	75	-
2855N	Orchard/Vineyard	1987–1992		80	-
2955	Utility	1987–1992		85	1606
3155	Utility	1988–1992		95	1589
3255	Utility	1991–1993		100	OECD 1346
4055	Row-Crop	1988–1992		106	OECD 1617
4255	Row-Crop	1988–1992		120	OECD 1618
4455	Row-Crop	1988–1992		148	OECD 1619
4555	Row-Crop	1988–1992		162	OECD 1620
4755	Row-Crop	1989–1991		180	OECD 1621
4955	Row-Crop	1988–1992		210	OECD 1622

THE 60 SERIES

By Ryan Roossinck

The articulated 60 series tractors were a major departure from the articulated 50 Series. The familiar sheet metal was completely restyled for a much more modern, angular look. Power options ranged from 235 to 370 horsepower, and with multiple transmission options ranging from twelve to twenty-four speeds, Deere was pretty confident that it had something for every farmer!

The 60 series was more than just a fresh set of body panels and some different motor options, though. Advancements in the 8560 finally solved a problem that John Deere had been dealing with for over thirty years, creating a clearer field of view out the front windshield. The engineers moved the air cleaner under the hood, which allowed them to then re-route the exhaust off to the right side on the 8560. It also allowed the use of a one-piece upper windshield as well, and that made a dramatic improvement in the operator's ability to see!

It trickled down to the large-frame row-crop tractors too. Although there was no increase in horsepower, there were enough improvements to the 4555, 4755, and 4955 to justify the model change. The row-crop tractors got the updated engine with the exhaust on the right, a bigger operator's platform, and re-engineered adjustable steps with two handrails (known today as "new style" steps, and often retrofitted to previous generation Sound-Gard tractors). Additionally, new lighting options made it easier than ever before to operate at night—especially in close proximity when unloading combines during harvest!

60 SERIES DATA				
Model	Type	Model Years	HP	Nebraska Test #
4560	Row-Crop	1992–1994	162	1620
4760	Row-Crop	1992–1994	180	1621
4960	Row-Crop	1992–1994	210	1622
8560	4WD	1989–1993	210	OECD 1623
8760	4WD	1989–1993	277	OECD 1624
8960	4WD	1990–1994	337	OECD 1625

JOHN DEERE XR91

These concepts from HDA show the development of the all-new John Deere line of four-wheel-drive tractors. The white models are from January 3, 1984. The sketch and prototype are not dated. *Henry Dreyfuss Archive, Cooper Hewitt, Smithsonian Design Museum*

1991 MODEL 8960

An all-new articulated four-wheel-drive line was introduced in Denver in the fall of 1988. The dream of the HDA designers from the early 1970s finally saw a longer wheelbase, which helped smooth the ride. *Renner Collection / Lee Klancher*

4X60 SERIES

For 1992, John Deere updated the larger 4X55 series with its revised 4560, 4760, and 4960. With the new thousand series models not far off, the update was fairly minor. *Chad Colby*

1993 MODEL 8970

For 1993, the big four-wheel-drive line received an update featuring more horsepower. The 8970 was up to a gross rating for 400 horsepower, and a base price of $144,175. *Renner Collection / Lee Klancher*

THE 70 SERIES

By Ryan Roossinck

For more than twenty years, the Sound-Gard body had set the standard for operator comfort in the field, and really redefined the farmer's expectation of what tractors should be. All things considered, the Sound-Gard tractors had a great run. The innovations that Deere introduced with the 30 series coupled with several decades of continuous improvement ushered in a new era of farming.

In 1993, the 70 series Power Plus tractors—the last of the aging Sound-Gard design—were released. The overall focus for these big articulated work-horses was efficiency. They were designed to be fairly low-maintenance, overbuilt powerhouses ready for long days in the field. The 8570, 8770, all-new 8870, and the monster 8970 were built in Waterloo from 1993 to 1996, until they were replaced by the 9000 series.

Mechanically, they were similar to the previous 60 series tractors, with a modest power bump. Deere also implemented one fairly major enhancement, the electronic throttle. It was implemented across the 70 series lineup to provide a more linear torque curve across the powerband. It also offered an on-demand power boost for situations where it was necessary.

In the cab, the big tractors continued with Deere's original mindset when developing the Sound-Gard body: make long hours in the cab more comfortable. The 70 series greenhouse was roomier than ever and was built with upgraded sound deadening materials, so it was quieter than ever, too. Coupled with a fully adjustable air ride seat and in-cab storage to reduce clutter, it made for a very comfortable mobile office!

These tractors are still hard at work in fields and construction sites all over the country today, which speaks volumes to Deere's commitment to quality.

1993 MODEL 8970

Three transmission choices were offered: the twelve-speed partially synchronized; a twenty-four-speed, two-speed Power Shift; and a twelve-speed full Power Shift. Fuel capacity was 220 gallons.
Renner Collection /
Lee Klancher

70 SERIES DATA

Model	Type	Model Years	HP	Nebraska Test #
8570	4WD	1993–1996	208	OECD 1670
8770	4WD	1993–1996	259	OECD 1671
8870	4WD	1993–1996	303	OECD 1672
8970	4WD	1993–1996	354	OECD 1673

CHAPTER SEVEN

21ST-CENTURY DEERE

MODERN GREEN TRACTORS

"We may see the farmer of the future moving a few feet above the ground in a sort of hovering jeep, electronically loosening the dirt, fertilizing it, and seeding it from the air. Closed-circuit TV may allow him to observe these fields from his living room."
—Henry Dreyfuss, Luncheon talk to Society of the Plastics Industry, Chicago, February 6, 1958

MODEL 8320

By the turn of the twenty-first century, the row-crop tractor had come of age. This model's 8.1-liter diesel cranks out 248 PTO horsepower, is four-wheel-drive, and the rear hydraulic lift capacity is 13,730 pounds. *Lee Klancher*

As Deere entered the mid-1990s, its position could not have been much better. After Robert A. Hanson led Deere through the tough times of the 1980s with careful investment and measured product launches, Hans W. Becherer was elected chairman in 1990 and was able to continue the company's torrid growth that had begun in the late 1950s.

Three decades of aggressive research and development had put distance between itself and the competition. By 1998, a report on John Deere Agricultural Operations by David Ocanas et al. showed Deere as the world's second largest company in the agricultural and construction business worldwide, with $13.8 billion in annual revenue.

Caterpillar held the top position with $20.9 billion, but its business is largely construction. The nearest agricultural competitor, Case, was in the third slot at $6.1 billion—less than half the revenue of Deere.

The challenges in the market at that time were considerable. Farmers continued to need to work more acres with less help, as more and more people moved from rural areas to urban centers. While the machines did get more expensive and profitable, the number sold was reducing domestically and good international sales networks were required to amortize research and development costs.

Focusing on global markets carries the risk of slipping domestically. While the competition was significantly behind, Deere was aggressively pursuing market share. By the mid-1990s, Case IH had begun to recover from the difficult merger of International Harvester's agricultural division with J. I. Case. It had introduced the Magnum tractor in 1987, with an eighteen-speed full powershift transmission. The Magnum closed the gap and signaled that Deere could not stop developing its core market of high-horsepower tractors designed for the U.S. market.

Deere responded vigorously, introducing its various thousand series tractors—with a new transmission—in the 1990s.

"In the late 1980s, Case IH introduced a new tractor to the tractor market with a forward moved cab, fuel tank behind the cab, and an eighteen-speed powershift transmission," said former John Deere engineer Richard Michael. "John Deere decided this was a market threat and decided to update the current tractor line to be introduced in 1992. This included a forward-moved cab for better access, rear and side located fuel tanks, a front frame to support the engine with the transmission separated from the engine. The early transmission design was done under my supervision."

The new transmission and the modular component design of the new machines were revolutionary and ensured that Deere maintained its lead.

The Deere core values remained true to those of the founder, and they started with providing high-quality machinery. The company continued to lead its segment in research and development expenses, spending $444 million in 1997.

As an aside, Deere continued to be aggressive with R&D and fiscally conservative with product launches. Research on the Deere rotary separation combine began in the 1950s and an entire line was developed in the late 1970s and early 1980s. The production version of the rotary combine technology finally saw light in August 1999, when Deere released the new STS line of combines.

While engineers are not always fans of this process, CFOs love the fact that new product launches are done with a sharp eye on the bottom line. Deere's profit margins and debt ratios lead the industry, and that careful management gave the company cash and reserves necessary to compete in the volatile heavy equipment market.

In addition to releasing new tractor lines, Deere offered the unfortunately named and brilliantly conceived "Dress Up a Deere Friend" campaign that added up-to-date comfort and conveniences to older 30, 40, and 50 series model tractors. New lighting systems, air conditioning, seats, electrical plugs and systems, sound systems, and hydraulic control valves were offered, and engine overhauls and replacements were also available to keep the older machines working reliably.

One of the threats faced by the stubbornly independent Deere was consolidation. When Case Corp. and New Holland N.V. merged in 2000 to become CNH Global N.V., the combined assets rivaled that of Deere. The new company would spend heavily on research and development to ensure that Deere had to work to retain its leadership position.

Deere responded by doubling its research and development budget from 1999's $458 million to just over $1 billion in 2010. It also more than doubled gross revenue from $11,751 million in 1999 to $26,005 million in 2010.

An investment report on Deere done by the University of Oregon Investment Group on May 30, 2012, showed that Deere dominated the U.S. market for agricultural and turf machines at that time, holding a 46.2 percent market share with the nearest competitor in the category, CNH Global N.V., holding only 11.6 percent of the market. This may be slightly misleading, as Deere dominates the turf market, and the competition is undoubtedly closer if you look at only the agricultural products.

On a global scale, Deere and CNH Global N.V. had very similar gross sales, with Deere at $32,013 million and CNH Global at $32,224 million. Both companies experienced ups and downs in the market chaos of the past decade, and Deere had stretched out a lead again by 2020, raking in $35,540 million to CNH's $24,285 million.

Deere has faced new challenges in modern times. Its restrictive policy of making it illegal to modify its electronic systems had raised ire with farmers and inspired a cottage industry of offshore development of technology that can hack into Deere's proprietary electronic management system.

The size of the farm market continues to shrink as the market's need and demand for high technology grows. Automation is of keen interest to farmers working massive plots with minimal help, and

MODEL 7800 CUTAWAY

This detailed cutaway shows off Deere's new-for-1992 7000 series. *Henry Dreyfuss Archive, Cooper Hewitt, Smithsonian Design Museum*

competing in that market will require not only high R&D costs but also careful marketing. Farmers have reacted with very mixed responses to the idea of an unmanned tractor on their land, and any errors by the machines that incur damage will undoubtedly face massive press and customer backlash.

As it stands today, Deere is the undisputed market leader. The fact that it was able to compete with a massive conglomerate in the early part of the twentieth century and emerge as the unquestioned market leader for the latter half is one of the great success stories in American business, much less agriculture.

THE 6000 SERIES

by Simon Henley from *Classic Tractor Magazine*

When the John Deere 6000 series was launched in 1992, to the untrained eye it was hard to identify any discernible differences in its physical appearance from any other brand. In fact, apart from its new styling, at a glance the 6000 series models appeared to conform to conventional design. But look closer and the differences became apparent.

When John Deere developed the 6000 and 7000 series tractors in the late 1980s, it developed a strong, flexible steel chassis that utilized modular components. One of the key engineers involved in the project was Michael Teich, who had joined John Deere in

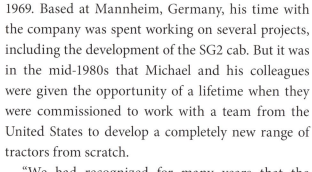

M3 SCALE MODEL

The M3 was the pre-production moniker for the 6200. This drawing was done by HDA.

Henry Dreyfuss Archive, Cooper Hewitt, Smithsonian Design Museum

1969. Based at Mannheim, Germany, his time with the company was spent working on several projects, including the development of the SG2 cab. But it was in the mid-1980s that Michael and his colleagues were given the opportunity of a lifetime when they were commissioned to work with a team from the United States to develop a completely new range of tractors from scratch.

"We had recognized for many years that the traditional stress supporting casting used in tractor engines and transmissions was not the way forward," explains Michael. "With tractors of 100 horsepower becoming more common, the problems of supporting bigger, heavier equipment on the castings was becoming a concern for engineers who recognized the loads implied by heavier ploughs and cultivators were unsustainable. It was decided, therefore, that using a chassis with modular or individual components was the only way forward."

With the 4030 model range of large conventional tractors in production by the early 1970s, John Deere was already aware of the limitations associated with designing the engine as a stressed frame member on big tractors. In 1979, Deere's Product Engineering Center at Waterloo, Iowa, conducted a feasibility study to explore the potential for using a main frame chassis and modular components in tractors. It was a study that would lay the path for a development project that would become known as the M-Tractor.

By 1984, senior executives at John Deere's world headquarters at Moline, Illinois, had decided to take the mainframe concept to the next stage. To do this a combined team of engineers from Germany and the United States were commissioned to produce a study exploring the concept of mainframe tractor production and to make a design proposal. The study was completed the following year, and this resulted in the team getting the go-ahead to build a mock-up of its new design, the first of which was completed in December 1985.

"The original experiments we conducted with frame design were not successful," says Michael.

"Initially we tried to incorporate the same rigidity in the frame as there was in the castings, but we quickly realized this would not work. The castings were designed not to flex, and trying to replicate this in a frame design was implausible.

"It was, therefore, decided we needed to incorporate flexibility into the design of the frame so it would move. It was a similar principle to the one used by Gustave Eiffel and his engineers when they designed the Eiffel tower in Paris. The tower was designed to distribute the stress loads applied by the wind by flexing.

"Vertical, side, and torsional loads were important considerations while designing the new frame for the M-Tractor," adds Michael. "Research showed that the z-shape design was the most cost-effective way to produce a mainframe that would handle the loads we required. The problem we had then was that the tractor's components, such as the engine and gearbox, were now trapped inside the frame, making them harder to access."

The workload imposed on Michael Teich and his colleagues was intense. The team—that included five engineers from Mannheim (Advanced Intermediate Tractor) and four engineers from Waterloo (Advanced Row-crop Tractor)—spent endless hours researching, calculating, and designing the new main frame concept, traveling back and forth between their respective countries—on several occasions for weeks on end—during the first twelve months of the program.

"You must consider this was to be a huge change in the manufacturing process that affected every single aspect of the tractor's production," says Michael. "Total tractor production in the factories would have to be devoted to mainframe design. That meant changes in the production of the engines, transmissions, axles, cabs—every single working component of the tractor. It would mean retooling and completely redesigning the production lines and retraining the entire production staff. It was a colossal undertaking!"

It was quickly realized by the engineers that one of the benefits of using a heavy-gauge steel chassis was that the components themselves would no longer need to be load bearing. This meant, for example, that the casting thickness of the engine block and transmission could be significantly reduced, and that lighter materials, such as aluminum, could be incorporated into the component designs.

Because the tractor's mainframe design limited access to components from outside the tractor, it was decided from the outset that the design must incorporate modular components. This basically meant that the tractors would be constructed on a building block system, with components such as the engine, transmission, PTO, and hydraulics being grouped into separate units that were linked as they were assembled in the frame.

This system also allowed the production line to quickly make changes in specification or add optional equipment without significant delays or changes in

M3 PROTOTYPE

This late prototype became the Model 6200. *Henry Dreyfuss Archive, Cooper Hewitt, Smithsonian Design Museum* production routines. Customers could, for example, specify a transmission option—such as a creeper box—that could be easily dropped in as the tractor was being built. In short, with such new-found flexibility, the module system benefited both the customer and the manufacturer alike.

Initially, production of the new tractors in the U.S. would be limited to the larger six-cylinder models, a range that would in time become known as the 7000 series, while Deere's German tractor factory at Mannheim would be used to produce a range of smaller versions of the M-Tractor—later named the 6000 series—ranging from 75 to 100 horsepower and all powered by a new four-cylinder engine. The tractors that became the 6100, 6200, 6300, and 6400 models were known internally as the M1, M2, M3, and M4. Six-cylinder versions for production in Germany were introduced at a later date.

M4 PROTOTYPE

A 1980s prototype of the four-cylinder M4 (100 horsepower) that became the John Deere 6400 when it entered production in 1992. The new tractors were styled by Henry Dreyfuss Associates. *Classic Tractor Magazine*

With a new sharp look that used only the most subtle of curves to soften its appearance, the new range was styled by the famous Michigan-based industrial design house Henry Dreyfuss Associates, a company whose long standing relationship with John Deere had originally started in 1938. The new sheet metal design represented a complete contrast to the curvaceous panels that had clad Deere products since the early 1960s, yet it managed to retain a distinctive look that made the range instantly recognizable as a John Deere product.

By this time product and design engineers on both sides of the Atlantic were working furiously to develop new electrical, hydraulic, and mechanical operating systems for the new tractors. This included not only new engines, but new transmission options, new hydraulic specifications, a new PTO system, new four-wheel drive systems, a brand-new cab design, and new accessories such as fore-end loaders.

In the engine department, a new turbocharged 4.5-litre four-cylinder unit was developed for the smallest and largest of the four new models, while the two mid-range models utilized a 3.9-litre four-cylinder turbo unit similar to the one used in the outgoing 50 series tractors. All the engines were manufactured by John Deere at its dedicated engine factory at Saran in France.

To complement the new engines, a new range of transmissions was offered, the most significant of which was PowrQuad, a new twenty forward/twenty reverse or twenty-four forward/twenty-two reverse unit that offered four powershift speeds in each of five or six ranges and a 40 kilometer per hour road speed. A forward/reverse shuttle was also incorporated into the design. Another significant step was the introduction of multi-plate wet clutch, or "a clutch for life" as John Deere marketing men described it at the time.

In the hydraulic department, a new electronically controlled, closed-circuit, on-demand hydraulic system that remained dormant until activated was used to reduce unnecessary power consumption, while

external lift cylinders exclusively handled the rear linkage. A three-speed (540/540E/1000) electronically activated PTO system further improved the new tractor's versatility, all four of which were available in two-wheel drive or with mechanical four-wheel drive (MFWD).

The new MFWD system featured a limited-slip differential and a fifty-two-degree castor angle on the front axle for tighter turning, and it was activated by a simple rocker switch. Additionally, to make full use of the tractor's in-board disc brakes and improve braking safety, the four-wheel drive system was also activated when the operator pressed the (latched) brake pedals.

John Deere's success with its American Sound-Gard body and German-built SG2 cabs had been exceptional. However, the split window, curved-screen cab was beginning to show its age and market research showed customers wanted a new cab with a more conventional appearance. The result was what became known in Europe as the TechCenter, a new, quieter 72dB(A) two-door unit with greater glass area, 40 percent more internal space, a high-flow ventilation system with optional air-conditioning, and an optional passenger seat.

John Deere had done a lot of development with molded plastics during the 1980s and had used them extensively in the new range's interior, which was now trimmed in a new tan color scheme. The dash tilted with the steering wheel and an ergonomically designed console to the driver's right placed all the main controls—including the gear levers, the electronic linkage control, the throttle, the PTO control, and the external hydraulic levers—literally at the operator's finger tips.

One huge benefit of the new cab over the previous SG2 was that the front windshield was no longer split, and the exhaust was now mounted at the front, right-hand side of the cab. Additionally, the driver no longer had to enter the cab through the opening front window door to get his seat, and the new wide doors on both sides offered unrestricted access. It was, in

short, a revelation in cab design that turned the tide for both John Deere dealers and farmers alike.

Gordon Day, who is now John Deere's branch tactical marketing manager in the UK, was involved in the launch of the 6000 series tractors during 1992. They were initially unveiled to European dealers in Germany in September, before being launched to UK dealers and customers at Deere's Langar, Nottinghamshire, base the following month.

"It was the biggest launch event since the original 30 series," says Gordon, who demonstrated the new tractors to UK customers. "It was a huge occasion for us, particularly since there was a lot of anticipation from British dealers. In fact, I remember that some people cheered and clapped when they first saw the tractors.

"For many farmers it was simply the novelty that the tractor cab now had two doors. It was the first time we'd offered a two-door cab, and I distinctly

PRE-PRODUCTION 6200

A pre-production pilot build version of the John Deere 6200 (known internally as the M2) showing the construction of the roll-frame that formed the basis of the TechCenter cab.
Classic Tractor Magazine

MODEL 6200

On the modular frame concept used by the 6000 series tractors, the main welded steel frame or backbone carried the load, leaving the power train free to transmit the power. The 6200 was introduced in 1993, a year after the other 6000 series machines. *Henry Dreyfuss Archive, Cooper Hewitt, Smithsonian Design Museum*

MODEL 6400

Powered by a new 4.5-litre turbocharged four-cylinder engine, the 100-horsepower John Deere 6400 was the largest of the four new 6000 Series models introduced in 1992. *Henry Dreyfuss Archive, Cooper Hewitt, Smithsonian Design Museum*

remember for some time after the launch that people who would never previously have bought a Deere suddenly started talking to us. Not everyone liked it at first, though. Some fans of the old SG2 cab simply didn't approve.

"The 6000 series was an incredible achievement," adds Gordon. "The cab had been moved forward, which shortened the bonnet, and featured an exceptionally well-designed control layout. We had the new four-speed semi-powershift transmission, a completely new on-demand hydraulic system, and completely new electronic linkage controls. We also had the industry's first push-back pick up hitch, which none of our competitors offered."

The new 6000 series modular design had also been developed to reduce maintenance downtime. A tilting cab gave unlimited access to the tractor's major components, making splitting rails—once a necessity to split the tractor in half—completely obsolete. Additionally, the tractors also featured a lift-up bonnet with easy access to filters and coolers, while every single hydraulic valve, including those for the hydraulic system and transmission, were now mounted externally under the cab.

"The 6000 series changed the game for John Deere in the UK," confesses Gordon. "We now had a range of tractors that appealed to everyone—a tractor range that incorporated features that none of our competitors had. It was undoubtedly a great time to be involved in the tractor industry."

The four new tractors—the 6100 (75 horsepower), 6200 (84 horsepower), 6300 (90 horsepower), and the 6400 (100 horsepower)—were an instant success, and John Deere quickly followed them up the following year with two six-cylinder models, the 6600 (100 horsepower) powered by a 5.9-litre engine and the 6800 (120 horsepower) powered by a new 6.8-litre unit. A third six-cylinder tractor, the 6900 (130 horsepower), powered by a more powerful version of the 6.8-litre engine, arrived in 1994.

"Customer feedback quickly revealed that the new range had been well accepted," recalls former Mannheim engineer Michael Teich, who retired in 2008. "We could see the new range was increasing John Deere's market share. I think people were amazed to see there was actually a tractor with a frame.

"Looking back, I believe using the modular design was an incredible vision," he adds. "It was the highlight of my career, being able to design a tractor from a clean sheet of paper. I personally had a chance as an engineer to be involved in one of the biggest innovations in tractor design. It was a remarkable experience."

Today it's the newly launched 6R Series that continues John Deere's M-Tractor legacy. These latest high-tech models represent a twenty-year succession of more than half a million tractors.

2000, 5000, & 6000 SERIES DATA

Model	Type	Model Years	Notes	HP	Nebraska Test #
2000	Utility	1993–1998	Built in Brno, Czech Republic.	45	-
2100	Utility	1993–1998	Built in Brno, Czech Republic.	57	-
2200	Utility	1993–1998	Built in Brno, Czech Republic.	63	-
2300	Utility	1993–1998	Built in Brno, Czech Republic.	69	-
2400	Utility	1993–1998	Built in Brno, Czech Republic.	78	-
2700	Utility	1993–1998	Built in Brno, Czech Republic.	82	-
2800	Utility	1993–1998	Built in Brno, Czech Republic.	92	-
2900	Utility	1993–1998	Built in Brno, Czech Republic.	103	-
5200	Utility	1992–1997	Built in Augusta, Georgia.	42	1660
5300	Utility	1992–1997	Built in Augusta, Georgia.	51	1661
5400	Utility	1992–1997	5400N 1995-1997; built in Augusta, Georgia.	63	1662
6100	Utility	1992–1997	6100SE Economy model; MFWD.	75	-
6200	Utility	1992–1997	6200SE Economy model.	84	-
6300	Utility	1992–1997	6200SE; 6300L Low-Profile.	90	S157
6400	Utility	1992–1998	6400SE; 6400L Low-Profile; 6400SP (Mexican market).	100	S158
6600	Row-Crop	1993–1997		110	-
6800	Row-Crop	1993–1997		120	-
6900	Row-Crop	1993–1996		130	-

THE 7000 SERIES

While the 6000 series was being developed, the Waterloo-based Deere team was busy creating the higher-horsepower 7000 series.

The new TechCenter cab debuted on the 7000 series with improved visibility and sound reduction, as well as digital display panels. The console was extensively redesigned and allowed easier control of the machine. The transmission was also the new nineteen-speed Power Shift (one more than the pesky competition's new powershift unit).

William Crookes was an industrial designer at Henry Dreyfuss and Associates (HDA) who worked on the John Deere account for many years. He said that while Bill Hewitt was CEO at Deere, the communication between John Deere and HDA was outstanding, and that they were involved from the beginning to the end of the process. He said that the process was quite different by the time the 7000 series was developed.

When the 7000 series was conceived, the process began with a meeting at which the parameters were defined. Crookes said the engineering team had more say than in the past. "We would get the group together and define the givens. And when we did that, a lot of times it was how agreeable the Deere engineers would be to the next generation of appearance," Crookes said. "From time to time they would not agree on things like what finger protection you need downstream over the fan, the cooling fan.

"They also changed CEOs, so we did not have a line of communication, and neither did we have the PRMs that reviewed the design proposals and configurations. We lost that link and then it was up to

DESIGNING A NEW LINE

The new 7000 series was an all-new design penned by Henry Dreyfuss Associates. Dan Nichols and Ms. Fructus at HDA are shown here drawing on a new hood design. *Henry Dreyfuss Archive, Cooper Hewitt, Smithsonian Design Museum*

NEW 7000 SERIES

The new 7000 series featured a modular design that no longer used the engine as a stressed member of the frame. *Henry Dreyfuss Archive, Cooper Hewitt, Smithsonian Design Museum*

MODEL 7600

The pivoting front axle is featured in this shot of the smallest member of the line. *Henry Dreyfuss Archive, Cooper Hewitt, Smithsonian Design Museum*

MODEL 7700

The 7000 series machines were available with a mechanical front-wheel-drive, locking rear differential, and sixteen-speed partial Power Shift or nineteen-speed full Power Shift transmissions. *Henry Dreyfuss Archive, Cooper Hewitt, Smithsonian Design Museum*

TECHCENTER ENCLOSURE

The TechCenter cab had two doors, a tilting and telescoping steering wheel, and a noise level as a low as 72dB(A). *Henry Dreyfuss Archive, Cooper Hewitt, Smithsonian Design Museum*

whatever the powers within each design group to determine what was pertinent for that generation of tractor design."

Crookes suggested that this process allowed the continuity of design found on older series to start to fail. He said he saw that in the case of the 7000 series, when Ron Burk at Waterloo led a very ambitious redesign.

"There was a very handsome design . . . done by Ron Burk in Waterloo," Crookes said. "It centered around relocation of the cooling package. And ultimately it was produced and was well received, but not everybody could move their cooling package in the same way that the 7000 could. So that configuration was limited to the 7000s and couldn't be applied to the 8000s or the 9000s. So the whole design continuity started to fall apart.

"We got our arms twisted a number of times to be agreeable and let's do this. So we did it and ultimately wound up with this sort of discombobulated appearance of things."

MODEL 7800

The biggest model of the line was tested at Nebraska from May to July 1993. On most Deere tractors, the test results exceed the rating, and the 7800 was no exception. The tractor was rated for 145 PTO horsepower, and put out 159 in Nebraska test 1668. *Henry Dreyfuss Archive, Cooper Hewitt, Smithsonian Design Museum*

7000 SERIES DATA

Model	Type	Model Years	Notes	HP	Nebraska Tractor Test #
7600	Row-Crop	1992–1996	Built at Waterloo, Iowa.	113	1664
7700	Row-Crop	1993–1996	Built at Waterloo, Iowa.	126	1666
7800	Row-Crop	1992–1996	Built at Waterloo, Iowa.	158	1668
7200	Row-Crop	1993–1996	Built at Waterloo, Iowa.	98	1678
7400	Row-Crop	1993–1996	Built at Waterloo, Iowa.	105	1680

THE 8000 SERIES

NEW 8000 SERIES

Deere introduced a new line of high-horsepower row-crop tractors for the 1995 season, featuring the new CommandView cab, a fancy control center, a sixteen-speed full Power Shift transmission, optional row guidance, and standard 4×4 on the 8400. This is a 1995 8400. Three track-equipped models were added in 1997, prompting Caterpillar to promptly file a patent infringement lawsuit, stating the Deere design infringed on the Caterpillar Challenger track system. The courts ruled Deere did not infringe on Caterpillar's patents. *John Deere Archives*

8000 / 8000T SERIES DATA

Model	Type	Model Years	Notes	HP	Nebraska Tractor Test #
8100	Row-Crop	1995–1999	Built at Waterloo, Iowa.	179	OECD 1688
8200	Row-Crop	1995–1999	Fitted with the 8.1L engine for the 1997 model year, starting with tractor serial number 10001.	202	OECD 1689
8300	Row-Crop	1995–1999	Fitted with the 8.1L engine for the 1997 model year, starting with tractor serial number 10001.	225	OECD 1690
8400	Row-Crop	1995–1999	Built at Waterloo, Iowa.	252	OECD 1691
8200T	Agricultural Crawler	1997–1999	Built at Waterloo, Iowa.	202	OECD 1745
8300T	Agricultural Crawler	1997–1999	Built at Waterloo, Iowa.	226	OECD 1746
8400T	Agricultural Crawler	1997–1999	Built at Waterloo, Iowa.	255	OECD 1747

Machinery Pete covers auction prices and large tractor collections. He compiled this list of the most popular 1990s tractors, and the highest prices on 1990s machines.

TOP 5 1990S JOHN DEERE TRACTOR MODELS

1. John Deere 6400
2. John Deere 4960
3. John Deere 7800
4. John Deere 4455
5. John Deere 7810

· These five models are the most popular tractor searches April–June 2021 at www.MachineryPete.com

TOP 4 RECENT AUCTION PRICES ON 1990S MODEL DEERE TRACTORS

1. 1992 John Deere 4760 with 3 hours, sold: $197,028 on 10/28/20 Saskatchewan auction (Record price by $111,528)
2. 1998 John Deere 8200 with 1,490 hours, sold: $110,500 on 4/10/21 Nebraska auction (Record price by $18,000)
3. 1993 John Deere 4960 with 2,499 hours, sold: $93,250 on 2/9/21 Indiana auction (Highest auction price on 4960 in eight years—third highest auction price ever)
4. 1993 John Deere 7800 2WD with 360 hours, sold: $86,000 on 7/25/20 Missouri auction (Record price by $5,000)

THE 9000 SERIES

9000 SERIES DATA

Model	Type	Model Years	Notes	HP	Nebraska Tractor Test #
9200	4WD	1996–2001	Built at Waterloo, Iowa.	297	OECD 1731 OECD 1789
9100	4WD	1997–2001	Built at Waterloo, Iowa.	249	OECD 1730
9300	4WD	1997–2001	Built at Waterloo, Iowa.	302	-
9400	4WD	1997–2001	Built at Waterloo, Iowa.	312	OECD 1773 (also 9420)
9300T	Agricultural Crawler	2000–2001	Built at Waterloo, Iowa.	346	OECD 1790
9400T	Agricultural Crawler	2000–2001	Built at Waterloo, Iowa.	347	OECD 1791 (also 9420T)

NEW 9000 SERIES

The 70 series high-horsepower, four-wheel-drive line was replaced when the 9000 series debuted for the 1996 model year. The big machines had up to 312 horsepower, offering the big acreage farmer an all-new machine. The 9300T and 9400T tracked versions were added in 2000. *Marcus Pasveer*

THE X010 SERIES

MODEL 7810

Introduced along with the new 9000 series was a revised 7000 series and some new 5000 series machines as well. *Henry Dreyfuss Archive, Cooper Hewitt, Smithsonian Design Museum*

MODEL 6210

New John Deere X010 series (also called TEN series) machines came out throughout the late 1990s. This 6210 was introduced for the 1999 model year. *Machinery Pete*

X010 SERIES DATA

Model	Type	Model Years	Notes	HP	Nebraska Tractor Test #
5010 Series					
5210	Utility	1998–2001	Built at Augusta, Georgia.	46	1754
5310	Utility	1998–2001	Built at Augusta, Georgia.	59	1755
5410	Utility	1998–2001	Built at Augusta, Georgia.	66	1756
5510	Utility	1998–2001	Built at Augusta, Georgia.	82	1757
6010 Series					
6510	Row-Crop	1997–2001	6510L Orchard/Vineyard 1999–2002.	92	-
6610	Row-Crop	1997–2001		101	-
6810	Row-Crop	1997–2001	6810S high-speed.	113	-
6910	Row-Crop	1997–2001		125	-
6110	Utility	1999–2002	6110L Low-Profile; Mannheim, Germany.	71	OECD 1833
6210	Utility	1999–2002	6210L Low-Profile; 6210S Low-Profile cab; Mannheim, Germany.	79	OECD 1821
6310	Utility	1999–2002	6310L Low-Profile ROPS; 6310S Low-Profile cab; Mannheim, Germany.	87	OECD 1807
6410	Utility	1999–2002	6410L Low-Profile; 6410S Low-Profile cab; Mannheim, Germany.	98	OECD 1806
6010SE	Utility	1998–2003	4×2, 4×4 MFWD.	75	-
6110SE	Utility	1998–2003	4×2, 4×4 MFWD.	80	-
6210SE	Utility	1998–2003	4×2, 4×4 MFWD.	90	-
6310SE	Utility	1998–2003	4×2, 4×4 MFWD.	100	-
6410SE	Utility	1998–2003	4×2, 4×4 MFWD.	105	-
7210	Row-Crop	1997–2002	7210 Hi-Crop; built at Waterloo, Iowa.	98	OECD 1738
7410	Row-Crop	1997–2002	7410 Hi-Crop; built at Waterloo, Iowa.	110	OECD 1740
7610	Row-Crop	1997–2002	Built at Waterloo, Iowa.	127	OECD 1724
7710	Row-Crop	1997–2003	Built at Waterloo, Iowa.	152	OECD 1726
7810	Row-Crop	1997–2003	Built at Waterloo, Iowa.	168	OECD 1728
7510	Row-Crop	1999–2002	7510 Hi-Crop; built at Waterloo, Iowa.	118	OECD 1786
8010 Series					
8110	Row-Crop	1999–2002	Built at Waterloo, Iowa.	165	OECD 1772
8210	Row-Crop	1999–2002	Built at Waterloo, Iowa.	185	OECD 1773
8310	Row-Crop	1999–2002	Built at Waterloo, Iowa.	205	OECD 1775
8410	Row-Crop	1999–2002	Built at Waterloo, Iowa.	270	OECD 1777
8110T	Agricultural Crawler	2000–2001	Built at Waterloo, Iowa.	165	-
8210T	Agricultural Crawler	2000–2001	Built at Waterloo, Iowa.	216	OECD 1774
8310T	Agricultural Crawler	2000–2001	Built at Waterloo, Iowa.	235	OECD 1776
8410T	Agricultural Crawler	2000–2001	Built at Waterloo, Iowa.	271	OECD 1778
Advantage Series					
6405	Utility	1998–2002	Mannheim, Germany.	85	-
6605	Utility	1998–2002	Mannheim, Germany.	95	-
7405	Utility	1998–2002	Monterrey, Mexico.	105	-
5105	Utility	2000–2007	Built at Augusta, Georgia.	48	1770
5205	Utility	2000–2007	Built at Augusta, Georgia.	52	1771

THE X020 SERIES

One of the ongoing challenges of engineering new farm machinery in the twenty-first century is meeting the increasingly stringent U.S. Environmental Protection Agency (EPA) emissions. For manufacturers, the trick is to reduce emissions *and* offer improvements in fuel economy and power output. When John Deere's engineering team looked to upgrade the X010 machines, the challenge of matching EPA's Tier II emissions provided an opportunity. Retired Deere engineer Ron Burk explained.

"The components that had to be added for Emissions Regulations compliance drove the cost to build the tractor up but the customer saw no value in meeting Regulations, so we were forced to stretch the power up so we could increase the customer value that way."

The two key components that needed development were the cooling system and the frame. Meeting the new requirements required better cooling, and Burk and his team incorporated an air-to-air charge cooler to reduce combustion temperatures, which in turn reduced the formation of NOx (which the EPA regulations required they reduce).

"Second, these models also incorporated an exhaust gas recirculation (EGR) cooler. This allowed them to meet the regulations without a fuel economy penalty," Burk said. "Allowing the cooling package to have more frontal area reduced the power needed to operate the fan at full load, another feature that helped fuel economy."

The radiators were moved in front of the front tires to improve airflow. And Burk noted that by engineering these improvements, they were able to gain market share. Competitors who met the regulations without improving fuel economy lost sales.

While the goal had been to keep the styling of previous models, the new cooling package required more space than the old design would allow. For the new design, Burk worked closely with long-time HDA industrial designer William "Bill" Crookes.

In order to change the design, more strength was required. The steel frame of the older model was replaced with a new cast iron frame. "It was a little heavier than the steel one it replaced, but much stronger," Burk said. "Cast iron gave us the opportunity to give it a more sculpted shape to clear the tires. In my opinion, it looked better too—see how it complements the hood. As Bill Crookes is fond of saying, 'Form follows function!'

"We designed the cooling package first to *function*. Then we turned the design for the hood over to Bill to create a form that complemented that function.

"We gave him an additional challenge. We asked him not to create something that looked like an evolution of previous models. We asked instead for a form that was as different as the new tractor was from the one it replaced, and different from other tractors we were building. (We really wanted this new model to set a new standard for appearance.)"

Burk noted that the prototypes were built during one of the most significant moments of the past twenty years.

"We were building the prototypes for the 7020s in September of 2001. Prototype builds are the height of tension in a new tractor program. We were as tense as anyone could be trying to round up all the late delivered parts to build these tractors. The hood was one of those critical late parts. Then the world stopped on the morning of September 11, 2001.

"My supervisor got me on the phone, 'Stop what you are doing. Find every person who reports to you to make sure they are safe, and report back to me.'"

His team was safe, and the end result of the hard work of the Deere engineering team and the HDA design group led by William Crookes is a striking machine all of them took great pride in creating.

"We knew we had a winner when we showed the early prototypes to farmer customers, and one of them said, 'I know this sound silly, but I want to buy one of these and park it in my front yard with a spotlight on it for the neighbors to see.'"

MODEL 7920

The large-frame X020 series tractors received a significant redesign that created these handsome machines. Deere engineer Ron Burk and HDA industrial designer William "Bill" Crookes worked together on the design, which first appeared on the 7720 and 7820 as well as the 7920. "It would be hard to decide who was more proud of how it turned out, Bill, or me," Burk said. *Machinery Pete*

MODEL 8420

The handsome 7X20 design was later incorporated into the 8420. *Machinery Pete*

MODEL 9620T

Three high-horsepower models, the 9320T, 9420T, and 9620T, were offered with tracks. *Chad Colby*

MODEL 9320

In 2001 and 2002, Deere began rolling out its new Twenty series machines, ranging from the 46-horsepower 5220 to the 378-horsepower 9320 shown here. The new infinitely variable transmission (IVT) was rolled out in 2002. *Jim Allen Collection*

X020/25 SERIES DATA

Model	Type	Model Years	Notes	HP	Nebraska Tractor Test #
6025 Series					
6225	Utility	2000–2016	Built at Mannheim, Germany.	78	-
6325	Utility	2000–2016	Built at Mannheim, Germany.	88	-
6425	Utility	2000–2016	Built at Mannheim, Germany.	98	-
6525	Utility	2000–2016	Built at Mannheim, Germany.	102	-
5020 Series					
5220	Utility	2002–2004	Built at Augusta, Georgia.	46	1754
5320	Utility	2002–2004	Built at Augusta, Georgia.	58	1755
5420	Utility	2002–2004	Built at Augusta, Georgia.	67	1756
5520	Utility	2002–2004	Built at Augusta, Georgia.	82	1757
6020 Series					
6520 Premium	Utility	2001–2006	Models updated in 2003.	97	-
6620 Premium	Utility	2001–2006	Models updated in 2003.	110	-
6820 Premium	Utility	2001–2006	Models updated in 2003.	122	-
6920 Premium	Utility	2001–2006	Models updated in 2003.	134	-
6920S Premium	Utility	2001–2006	Models updated in 2003.	141	-
7020 Series					
7220	Row-Crop	2003–2007	Built at Waterloo, Iowa.	104	OECD 1816
7320	Row-Crop	2003–2007	Built at Waterloo, Iowa.	117	OECD 2038
7420	Row-Crop	2003–2007	7420HC Hi-Crop; built at Waterloo, Iowa.	127	OECD 1817
7520	Row-Crop	2003–2007	Built at Waterloo, Iowa.	144	OECD 1818
7720	Row-Crop	2003–2007	Built at Waterloo, Iowa.	160	OECD 1833
7820	Row-Crop	2003–2007	Built at Waterloo, Iowa.	174	OECD 1834 OECD 1834A
7920	Row-Crop	2003–2007	Built at Waterloo, Iowa.	193	OECD 1835
8020 Series					
8120	Row-Crop	2002–2005	Built at Waterloo, Iowa.	200	OECD 1819
8120T	Agricultural Crawler	2002–2005	Built at Waterloo, Iowa.	170	-
8220	Row-Crop	2002–2005	Built at Waterloo, Iowa.	222	OECD 1820
8220T	Agricultural Crawler	2002–2005	Built at Waterloo, Iowa.	216	OECD 1774
8320	Row-Crop	2002–2005	Built at Waterloo, Iowa.	248	OECD 1798
8320T	Agricultural Crawler	2002–2005	Built at Waterloo, Iowa.	235	OECD 1799
8420	Row-Crop	2002–2005	Built at Waterloo, Iowa.	269	OECD 1800
8420T	Agricultural Crawler	2002–2005	Built at Waterloo, Iowa.	271	OECD 1778 (Also 8410T)
8520	Row-Crop	2002–2005	Built at Waterloo, Iowa.	293	OECD 1801
8520T	Agricultural Crawler	2002–2005	Built at Waterloo, Iowa.	292	OECD 1802
9020 Series					
9120	4WD	2002–2007	Built at Waterloo, Iowa.	245	OECD 1810
9220	4WD	2002–2007	Built at Waterloo, Iowa.	334	OECD 1811
9320	4WD	2002–2007	Built at Waterloo, Iowa.	378	OECD 1803
9320T	Agricultural Crawler	2002–2007	Built at Waterloo, Iowa.	389	OECD 1804
9420	4WD	2002–2007	Built at Waterloo, Iowa.	312	OECD 1773 (also 9400)
9420T	Agricultural Crawler	2002–2007	Built at Waterloo, Iowa.	347	OECD 1791 (also 9400T)
9520	4WD	2002–2007	Built at Waterloo, Iowa.	370	OECD 1805
9520T	Agricultural Crawler	2002–2007	Built at Waterloo, Iowa.	372	OECD 1806
9620T	Agricultural Crawler	2002–2007	Built at Waterloo, Iowa.	374	OECD 1845
9620	4WD	2004–2007	Built at Waterloo, Iowa.	377	OECD 1844

THE X030 SERIES

NEW HOODS

These hoods are prototypes shown at the HDA facilities in New York. *Henry Dreyfuss Archive, Cooper Hewitt, Smithsonian Design Museum*

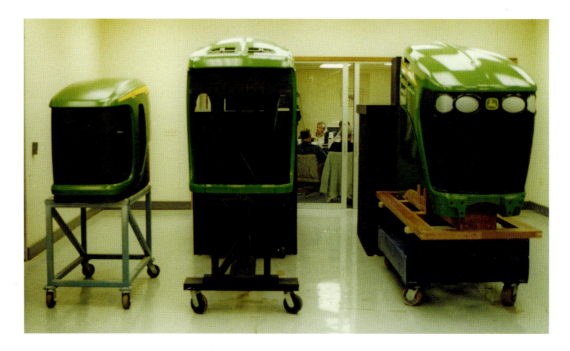

MODEL 9630

The new line of X030 series John Deeres were rolled out in 2006 and 2007. *Chad Colby*

X030 SERIES DATA

Model	Type	Model Years	Notes	HP	Nebraska Tractor Test #
6030 Series					
6230	Utility	2007–2012	Built at Mannheim, Germany.	75	-
6330	Utility	2007–2012	Built at Mannheim, Germany.	81	-
6430	Utility	2007–2012	Built at Mannheim, Germany.	111	1987
7030 Series					
7630	Row-Crop	2007–2011	Built at Waterloo, Iowa.	162	OECD 1893
7730	Row-Crop	2007–2011	Built at Waterloo, Iowa.	173	OECD 1894
7830	Row-Crop	2007–2011	Built at Waterloo, Iowa.	189	OECD 1895
7930	Row-Crop	2007–2011	Built at Waterloo, Iowa.	203	OECD 1897
7030 Small-Frame Series					
7130	Row-Crop	2007–2011	Mannheim, Germany; 7130 Premium; built at Waterloo, Iowa.	100	-
7230	Row-Crop	2007–2017	Mannheim, Germany; 7130 Premium; built at Waterloo, Iowa.	110	-
7330	Row-Crop	2007–2017	7330 Premium; 7330 Premium Hi-Crop; 2011 production moved from Waterloo, Iowa to Mannheim, Germany.	125	OECD 1952
7330 Premium	Row-Crop	2007–2011	Built at Waterloo, Iowa.	141	1952
7430 Premium	Row-Crop	2007–2011	Waterloo, Iowa and Mannheim, Germany.	161	OECD 1923
7530 Premium	Row-Crop	2007–2011	Introduced 2007, not available in U.S. until 2009; built in Waterloo, Iowa and Mannheim, Germany.	176	OECD 2453
8030 Series					
8130	Row-Crop	2006–2009	Built at Waterloo, Iowa.	208	OECD 1884
8230	Row-Crop	2006–2009	Built at Waterloo, Iowa.	233	OECD 1885
8230T	Agricultural Crawler	2006–2009	Built at Waterloo, Iowa.	233	OECD 1886
8330	Row-Crop	2006–2009	Built at Waterloo, Iowa.	261	OECD 1887
8330T	Agricultural Crawler	2006–2009	Built at Waterloo, Iowa.	270	OECD 1888
8430	Row-Crop	2006–2009	Built at Waterloo, Iowa.	289	OECD 1873
8430T	Agricultural Crawler	2006–2009	Built at Waterloo, Iowa.	298	OECD 1889
8530	Row-Crop	2006–2009	Built at Waterloo, Iowa.	312	OECD 1890
9030 Series					
9230	4WD	2007–2011	Built at Waterloo, Iowa.	312	OECD 1940
9330	4WD	2007–2011	Built at Waterloo, Iowa.	385	OECD 1941
9430	4WD	2007–2011	Built at Waterloo, Iowa.	383	OECD 1942
9430T	Agricultural Crawler	2007–2011	Built at Waterloo, Iowa.	368	OECD 1943
9530	4WD	2007–2011	Built at Waterloo, Iowa.	375	OECD 1924
9530T	Agricultural Crawler	2007–2011	Built at Waterloo, Iowa.	375	OECD 1925
9630	4WD	2007–2011	Built at Waterloo, Iowa.	375	OECD 1926
9630T	Agricultural Crawler	2007–2011	Built at Waterloo, Iowa.	375	OECD 1927

5M SERIES DATA

Model	Type	Model Years	Notes	HP	Nebraska Tractor Test #
5M Series					
5095M	Utility	2008–2011	5095MH high-clearance; built at Augusta, Georgia.	84	1961
5105M	Utility	2008–2011	Built at Augusta, Georgia.	82	1962
5065M	Utility	2008–2012	Built at Augusta, Georgia.	57	8195
5075M	Utility	2008–2017	Built at Augusta, Georgia.	66	1959
5085M	Utility	2008–2020	Built at Augusta, Georgia.	74	1960
5070M	Utility	2009–2013		70	-
5080M	Utility	2009–2013		80	-
5090M	Utility	2009–2013		90	-
5100M	Utility	2009–2013	Built at Augusta, Georgia.	84	OECD 2037 OECD 2116 OECD 2116A
5100MH	High-Clearance	2012–2017		100	-
5100ML	Orchard/Vineyard	2012–2017		100	-
5115M	Utility	2012–2017		103	OECD 2038 OECD 2117 OECD 2117A
5115ML	Orchard/Vineyard	2012–2017		116	-

THE R SERIES

NEW R SERIES

For the 2009 model year, John Deere introduced its 8R line of high-horsepower row-crop tractors. An 8335R is shown. The nomenclature is first digit is the series (8), the second three digits are the horsepower (335, in this image), and the last letter indicates type (R for row-crop). *Shelby Chesnut / GrandValeGalery dot com*

MODEL 8310R

The R series saw a host of new models introduced in 2011. Powered by dual-turbocharged 548-cubic-inch six-cylinder diesel, the 8310R could be had with sixteen-speed Automatic Power Shift transmission or an infinitely variable transmission (IVT). *Chad Colby*

MODEL 8360RT

Deere continued to offer most of its large tractors with a two-track system. While the early versions of this track system had a rough ride, Deere engineering improved matters significantly with the R series two-track machines. *Chad Colby*

MODEL 8335R

Another new-for-2011 model was the 8335R, which had a base price of $271,495. Fuel capacity was 184 gallons. *Shelby Chesnut / GrandValeGalery dot com*

9R SERIES

The new 9R high-horsepower, four-wheel-drive line debuted in 2012, offering gross engine ratings from 360 to 510 horsepower. Announced in August 2011, the machine was part of a significant product introduction that includes the new 6R series, the new 9R and 9RT, new S-series combines, and more. *Chad Colby*

9R SERIES UPDATE

In 2015, the 9R series was updated with higher horsepower and hydraulic capacity and improved hydraulics and transmissions. *John Deere Archives*

RX SERIES DEBUT

John Deere put its tracked machines on four tracks rather than two in 2016 when the new RX models appeared. This is a 9570RX. Four tracks offer an improved ride and better traction. *Chad Colby*

8R SERIES UPGRADES

The 8R series received some attention in 2016, with several new models added to the line. The 8400R added a higher-horsepower row-crop option, along with improved steering and more. *Chad Colby*

NEW NAMES AND FACES

The entire R series had an update to nomenclature, with the models named by series and horsepower separated by a space. Note the crisp new lines on the front of the 8R 310. Since 2012, John Deere has been contracting with Designworks, a BMW company, for industrial design work. Deere also has an in-house staff that does design. *Chad Colby*

R SERIES DATA

Model	Type	Model Years	Notes	HP	Nebraska Tractor Test #
5R Series					
5080R	Utility	2009–2016	Mannheim, Germany.	67	-
5090R	Utility	2009–2018	Mannheim, Germany.	78	-
5090R	Utility	2017–2021	Built at Augusta, Georgia.	90	-
5100R	Utility	2009–2020	Built at Augusta, Georgia.	110	OECD 2193 (open center hydraulic) OECD 2194 (pressure flow hydraulic)
5115R	Utility	2017–	Built at Augusta, Georgia.	115	OECD 2195 (open center hydraulic) OECD 2196 (pressure flow hydraulic)
5125R	Utility	2017–	Built at Augusta, Georgia.	125	OECD 2197 (open center hydraulic) OECD 2198 (pressure flow hydraulic)
6R Series					
6140R	Row-Crop	2012–2016	(2WD 4WD); built at Waterloo, Iowa.	134	OECD 2059
6150R	Row-Crop	2012–2016	(2WD 4WD) 6150RH Hi-Crop; built at Waterloo, Iowa.	145	OECD 2060
6170R	Row-Crop	2012–2014	(2WD 4WD); built at Waterloo, Iowa.	163	OECD 2033
6190R	Row-Crop	2012–2014	Built at Waterloo, Iowa.	189	OECD 2034
6210R	Row-Crop	2012–2014	Built at Waterloo, Iowa.	205	OECD 2035
6145R	Row-Crop	2015–2021	(2WD 4WD); built at Waterloo, Iowa.	134	OECD 2146
6155R	Row-Crop	2015–	Built at Waterloo, Iowa.	152	OECD 2147

Model	Type	Model Years	Notes	HP	Nebraska Tractor Test #
6175R	Row-Crop	2015–	Built at Waterloo, Iowa.	175	OECD 2136
6195R	Row-Crop	2015–	Built at Waterloo, Iowa.	189	OECD 2201
6215R	Row-Crop	2015–	Built at Waterloo, Iowa.	206	OECD 2106
7R Series					
7200R	Row-Crop	2011–2013	Built at Waterloo, Iowa.	197	OECD 2021
7215R	Row-Crop	2011–2013	Built at Waterloo, Iowa.	216	OECD 2005
7230R	Row-Crop	2011–2019	Built at Waterloo, Iowa.	225	OECD 2090 OECD 2022
7260R	Row-Crop	2011–2013	Built at Waterloo, Iowa.	257	OECD 2023
7280R	Row-Crop	2011–2013	Built at Waterloo, Iowa.	276	OECD 2024
7210R	Row-Crop	2014–2019	Built at Waterloo, Iowa.	170	OECD 2082 OECD 2179
7250R	Row-Crop	2014–2019	Built at Waterloo, Iowa.	205	OECD 2085
7270R	Row-Crop	2014–2019	Built at Waterloo, Iowa.	224	OECD 2084
7290R	Row-Crop	2014–2019	Built at Waterloo, Iowa.	283	OECD 2091
7310R	Row-Crop	2015–2019	Built at Waterloo, Iowa.	310	OECD 2113
7R 210	Row-Crop	2019–	Built at Waterloo, Iowa.	170	-
7R 230	Row-Crop	2019–	Built at Waterloo, Iowa.	189	-
7R 250	Row-Crop	2019–	Built at Waterloo, Iowa.	205	-
7R 270	Row-Crop	2019–	Built at Waterloo, Iowa.	224	-
7R 290	Row-Crop	2019–	Built at Waterloo, Iowa.	242	-
7R 310	Row-Crop	2019–	Built at Waterloo, Iowa.	260	-
7R 330	Row-Crop	2019–	Built at Waterloo, Iowa.	260	-
7R 350	Row-Crop	2021–	Built at Waterloo, Iowa.	260	-
8R Series					
8225R	Row-Crop	2009–2010	Built at Waterloo, Iowa.	221	OECD 1966
8245R	Row-Crop	2009–2010	Built at Waterloo, Iowa.	235	OECD 1967
8270R	Row-Crop	2009–2010	Built at Waterloo, Iowa.	257	OECD 1968
8295R	Row-Crop	2009–2010	Built at Waterloo, Iowa.	281	OECD 1969
8320R	Row-Crop	2009–2010	Also 2014-2019; built at Waterloo, Iowa.	307	OECD 1963 OECD 2101
8320RT	Crawler	2014–2019	Built at Waterloo, Iowa.	297	OECD 1971 OECD 2129 (w/e23 transmission)
8345RT	Agricultural Crawler	2009–2010	Built at Waterloo, Iowa.	345	OCED 1973
8345R	Row-Crop	2009–2010	Built at Waterloo, Iowa.	325	OECD 1972
8235R	Row-Crop	2011–2013	Built at Waterloo, Iowa.	340	OECD 2001
8260R	Row-Crop	2011–2013	Built at Waterloo, Iowa.	263	OECD 2002
8285R	Row-Crop	2011–2013	Built at Waterloo, Iowa.	288	OECD 2003
8310R	Row-Crop	2011–2013	Built at Waterloo, Iowa.	303	OECD 2004 OECD 2004A
8310RT	Agricultural Crawler	2011–2013	Built at Waterloo, Iowa.	300	OECD 1998
8335R	Row-Crop	2011–2013	Built at Waterloo, Iowa.	340	OECD 1990 OECD 1990A

R SERIES DATA (CONTINUED)

Model	Type	Model Years	Notes	HP	Nebraska Tractor Test #
8335RT	Agricultural Crawler	2011–2013	Built at Waterloo, Iowa.	323	OECD 1999
8360R	Row-Crop	2011–2013	Built at Waterloo, Iowa.	355	OECD 1991 OECD 1991A
8360RT	Agricultural Crawler	2011–2013	Built at Waterloo, Iowa.	355	OECD 2000 OECD 2000A
8370R	Row-Crop	2014–2019	Built at Waterloo, Iowa.	375	OECD 2102 OECD 2131 (w/e23 transmission)
8370RT	Agricultural Crawler	2014–2019	Built at Waterloo, Iowa.	308	OECD 2132 (w/e23 transmission)
8400R	Row-Crop	2016–2019	Built at Waterloo, Iowa.	388	OECD 2172 (w/e23 transmission)
8R 230	Row-Crop	2019–	Built at Waterloo, Iowa.	182	-
8R 250	Row-Crop	2019–	Built at Waterloo, Iowa.	200	-
8R 280	Row-Crop	2019–	Built at Waterloo, Iowa.	228	-
8R 310	Row-Crop	2019–	8RT 310 two-track 8RX 310 four-track.	257	-
8R 340	Row-Crop	2019–	8RT 340 two-track 8RX 340 four-track.	284	-
8R 370	Row-Crop	2019–	8RT 370 two-track 8RX 370 four-track.	310	-
8R 410	Row-Crop	2019–	8RT 410 two-track 8RX 410 four-track.	310	-

9R Series

Model	Type	Model Years	Notes	HP	Nebraska Tractor Test #
9360R	4WD	2012–2014	Built at Waterloo, Iowa.	346	OECD 2030
9410R	4WD	2012–2014	Built at Waterloo, Iowa.	378	OECD 2039
9460R	4WD	2012–2014	Built at Waterloo, Iowa.	375	OECD 2040
9460RT	Agricultural Crawler	2012–2014	Built at Waterloo, Iowa.	360	OECD 2041
9510R	4WD	2012–2014	Built at Waterloo, Iowa.	411 (db)	OECD 2042
9510RT	Agricultural Crawler	2012–2014	Built at Waterloo, Iowa.	392 (db)	OECD 2043
9560R	4WD	2012–2014	Built at Waterloo, Iowa.	454 (db)	OECD 2044
9560RT	Agricultural Crawler	2012–2014	Built at Waterloo, Iowa.	433 (db)	OECD 2045
9370R	4WD	2015–	Built at Waterloo, Iowa.	370 (gross)	OECD 2107
9420R	4WD	2015–	Built at Waterloo, Iowa.	420 (gross)	OECD 2108
9470R	4WD	2015–	Built at Waterloo, Iowa.	470 (gross)	OECD 2109
9470RT	Agricultural Crawler	2015–	Built at Waterloo, Iowa.	470 (gross)	OECD 2110
9470RX	Articulated Crawler	2016–	Built at Waterloo, Iowa.	353 (db)	OECD 2170
9520R	4WD	2015–	Built at Waterloo, Iowa.	520 (gross)	OECD 2111
9520RT	Agricultural Crawler	2015–	Built at Waterloo, Iowa.	520 (gross)	OECD 2112
9520RX	Articulated Crawler	2016–	Built at Waterloo, Iowa.	398 (db)	OECD 2171
9570R	4WD	2015–	Built at Waterloo, Iowa.	570 (gross)	OECD 2133
9570RT	Agricultural Crawler	2015–	Built at Waterloo, Iowa.	570 (gross)	OECD 2134
9570RX	Articulated Crawler	2016–	Built at Waterloo, Iowa.	447 (db)	OECD 2173
9620R	4WD	2015–	Built at Waterloo, Iowa.	620 (gross)	OECD 2135
9620RX	Articulated Crawler	2016–	Built at Waterloo, Iowa.	620 (gross)	OECD 2174

INDEX

ACKNOWLEDGMENTS

A special thanks to Bruce Keller, who agreed to build the first studio for this project at his fantastic collection in Brillion, Wisconsin. Bruce is a joy to work with, and his vast knowledge of Deere history was invaluable. Thanks are also due to Tom and Jake Renner, Jack Purinton, and all the other collectors who took time to dig out, polish, and move their machinery for photographs.

Thanks to those who offered help at various archives and organizations: Emily Orr, curator at the Cooper Hewitt, Smithsonian Design Museum, was a delight and helped locate critical new materials; the always enthusiastic Charles "Chuck" Pelly, whose last-minute find of a pile of terrific slides and drawings were invaluable. Also to Arthur L. Carlson at the George C. Gordon Library at Worcester Polytechnic Institute; Sarah Detweiller at the University of Wisconsin–Green Bay; the crew at the University of Wisconsin Historical Society; Suzy Weeks at Mecum Auctions; author Russell Flinchum; and Brian Holst and Neil Dahlstrom at John Deere.

Thanks to the book's terrific contributors, Jim Volgarino, Ryan Roossinck, Chad Colby, and Greg "Machinery Pete" Peterson.

Thanks to the engineers and innovators who offered their time, archives, and memories: Harold Brock, Gerald Mortensen, William Crookes, Rick Heth, Dale Johnson, Dr. Glenn Kahle, Jon Kinzenbaw, Richard Michael, Bob Haight, John Frye, Mark Weaver, Bud Youle, Don Watt, Michael Teich, Gordon Day, Jon Lauter, John Mannisto, Ron Burk, Dick Hurban, and Steve Mitchell.

To video and lighting wizard Josh Kufahl, for his invaluable photo and video knowledge, work ethic, and sense of humor. To Ethan VerKuilen for a great photo assist at the Half Century of Progress Show 2019. Others who helped make the photos include Bill Kapinski at Image Studios; Philip Cousins and Eileen Healy at Chimera; James F. Bland, Kirk Tuck, and the rest of the ASMP crew in Austin; Joseph Holschuh; Marv, Tom, and Mike, for all their efforts washing, pushing, polishing, and building on the set.

One of the joys of this book was working with the great crew at the Half Century of Progress Show, particularly Russell Buhr, John Frederickson, Corky Vericker, with generous support from Max Armstrong.

I am so appreciative of the hard work by our crew at Octane Press, particularly, designer Tom Heffron, for his hard work making this book look great; editor Maria Edwards for her sharp editorial eye; Catherine Mandel for terrific work marketing and promoting the book; also to Faith Garcia; Alicia Reynolds; and Eileen Hughes.

Last but never least, a big thanks to Joan Hughes for all her support, suggestions, patience, and love.

STUDIO WORK

Many of the images in this book were made using a Chimera F2X 10 × 20-foot overhead light bank, with eight Speedotron 202VF heads powered by four Speedotron 4803 power supplies. Multiple ground-based lights provided fill and highlights. Several locations were used—this is the studio built in Hangar 2 on the Chanute Air Force Base at the 2019 Half Century of Progress Show. *Lee Klancher*

BIBLIOGRAPHY

Broehl, Wayne G. Jr. *John Deere's Company*. New York: Doubleday &
    Company, 1984.
Brown, Theo. *Theo Brown Diaries*. Massachusetts: Gordon Library,
    Worcester Polytechnic Institute, 1893–1971.
Cherry, Jack, ed. *Two-Cylinder Magazine*. Iowa: Two-Cylinder, multiple
    issues, 1985–2021.
Covich, Edith S. *Max*. Illinois: Stuart Brent, 1974.
Dahlstrom, Neil and Jeremy. *The John Deere Story*. Illinois:
    Northern Illinois University Press, 2005.
Deere & Company Annual Reports and SEC filings. Illinois:
    Deere & Company, 1956–2020.
Deere & Company Investment Thesis. Oregon: University of Oregon
    Investment Group, 2012.
Dreyfuss, Henry. *Designing for People*. New York: Simon and Schuster,
    1955.
Dunning, Lorry. *John Deere Tractor Data Book*. Minnesota: Motorbooks,
    1996.
Ellis, L.W. *The Economic Importance of the Farm Tractor*. ASME, 1910.
Flinchum, Russell. *Henry Dreyfuss: The Man in the Brown Suit*.
    New York: Rizzoli, 1997.
Gay, Larry. *Farm Tractors 1975–1995*. Michigan: ASAE, 1995.
Gray, R.B. *The Agricultural Tractor 1855–1950*. Michigan: ASAE, 1954.
Hain, Richard. *John Deere 10 Series New Generation Tractors*.
    Nebraska: Hain Publishing Inc, 2004.
———. *John Deere 20 Series New Generation Tractors*.
    Nebraska: Hain Publishing Inc, 2012.
———. *John Deere 20 Series Two-Cylinder Tractors*.
    Nebraska: Hain Publishing Inc, 2010.
———. *John Deere Hi-Crop Data Book*. Nebraska: Hain Publishing Inc,
    2012.
Hewitt, William A. *Hewitt Family History From 1947, Vol. One and Two*.
    California: Privately published, 1999.
Hobbs, J.R. *John Deere Unstyled Letter Series*. Nebraska:
    Hain Publishing Inc, 2000.
John Deere Tractors 1918–1994. Michigan: ASAE, 1994.
Josephson, H. B. "The design of a general-purpose tractor."
    Iowa: Iowa State College, 1925.
Kinzenbaw, Jon. *50 Years of Disruptive Innovation*. Iowa:
    Bantry Bay Publishing, 2015.
Larsen, Lester. *Farm Tractors 1950–1975*. Michigan: ASAE, 1981.
Leffingwell, Randy. *John Deere*. Minnesota. Motorbooks, 2004.
MacMillan, Don & Harrington, Roy. *John Deere Tractors and Equipment,
    Volume Two 1960–1990*. Michigan: ASAE, 1991.
MacMillan, Don. *John Tractor Legacy*. Minnesota: Voyageur Press, 2003.
Magee, David. *The John Deere Way*. New Jersey: John Wiley & Sons, 2005.
Miller, Merle L. *Designing the New Generation John Deere Tractors*.
    Michigan: ASAE, 1999.
Ocanas, David, et al. *John Deere & Co. Agricultural Operations*. 1998.
Profit, Bill. "The John Deere Rotary Combine," Nebraska:
    Green Magazine, November 1989.
Simpson, Peter. *Ultimate Tractor Power, Vol. 1*. East Yorkshire,
    United Kingdom: Japonica Press, 2000.
———. *Ultimate Tractor Power, Vol. II*. East Yorkshire, United Kingdom:
    Japonica Press, 2002.

Octane Press, Edition 1.1
January 2022
© 2021 by Lee Klancher

ISBN: 978-1-64234-008-2
LCCN: 2021931352

Design by Tom Heffron
Copyedited by Maria Edwards
Proofread by Meagan Smith

octanepress.com

Octane Press is based in Austin, Texas

Printed in China

On the cover: This John Deere 6030 is owned and was restored by
Brad and Casey Walk. It is the first example built, serial number 033000R.
Lee Klancher

On the frontispiece: The serial number plate from the first Model GP
built, serial number 200211, owned by Bruce and Walter Keller.
Lee Klancher

On the title page: Model 7520 owned by Tom and Jake Renner.
Lee Klancher

On the front endpaper: Battered new generation tractor. *Lee Klancher*

On the back endpaper: Tom Renner and his Wagner. *Lee Klancher*

On the back cover: Lee Klancher, Cooper Hewitt Smithsonian Museum,
Chuck Pelly Collection